Thomas Henry Huxley

Man's Place in Nature and other Anthropological Essays

Thomas Henry Huxley

Man's Place in Nature and other Anthropological Essays

ISBN/EAN: 9783337026851

Printed in Europe, USA, Canada, Australia, Japan

Cover: Foto ©berggeist007 / pixelio.de

More available books at **www.hansebooks.com**

MAN'S PLACE IN NATURE

AND OTHER
ANTHROPOLOGICAL ESSAYS

BY
THOMAS H. HUXLEY

NEW YORK
D. APPLETON AND COMPANY
1896

Authorized Edition.

PREFACE

I AM very well aware that the old are prone to regard their early performances with much more interest than their contemporaries of a younger generation are likely to take in them; moreover, I freely admit that my younger contemporaries might employ their time better than in perusing the three essays, written thirty-two years ago, which occupy the first place in this volume. This confession is the more needful, inasmuch as all the premisses of the argument set forth in "Man's Place in Nature" and most of the conclusions deduced from them, are now to be met with among other well-established and, indeed, elementary truths, in the text-books.

Paradoxical as the statement may seem, however, it is just because every well-informed student of biology ought to be tempted to throw these essays, and especially the second, "On the Relations of Man to the Lower Animals," aside, as a fair mathematician might dispense with the reperusal of Cocker's arithmetic, that I think it worth while to reprint them; and entertain the hope that the story of their origin and early fate may not be devoid of a certain antiquarian interest, even if it possess no other.

In 1854, it became my duty to teach the principles of biological science with especial reference to paleontology. The first result of addressing myself to the business I had taken in hand, was the discovery of my own lamentable ignorance in respect of many parts of the vast field of knowledge through which I had undertaken to guide others. The second result was a resolution to amend this state of things to the best of my ability; to which end, I surveyed the ground; and having made out what were the main positions to be captured, I came to the conclusion that I must try to carry them by concentrating all the energy I possessed upon each in turn. So I set to work to know something of my own knowledge of all the various disciplines included under the head of Biology; and to acquaint myself, at first hand, with the evidence for and against the extant solutions of the greater problems of that science. I have reason to believe that wise heads were shaken over my apparent divagations—now into the province of Physiology or Histology, now into that of Comparative Anatomy, of Development, of Zoology, of Paleontology, or of Ethnology. But even at this time, when I am, or ought to be, so much wiser, I really do not see that I could have done better. And my method had this great advantage; it involved the certainty that somebody would profit by my effort to teach properly. Whatever my hearers might do, I myself always learned something by lecturing. And to those who have experience of what a heart-breaking business teaching is—how much the can't-learns and won't-learns and don't-learns predominate over the do-learns—will understand the comfort of that reflection.

Among the many problems which came under my consideration, the position of the human species in zoological classification was one of the most serious. Indeed, at that time, it was a burning question in the sense that those who touched it were almost certain to burn their fingers severely. It was not so very long since my kind friend Sir William Lawrence, one of the ablest men whom I have known, had been well-nigh ostracized for his book "On Man," which now might be read in a Sunday-school without surprising anybody; it was only a few years, since the electors to the chair of Natural History in a famous northern university had refused to invite a very distinguished man to occupy it because he advocated the doctrine of the diversity of species of mankind, or what was called "polygeny." Even among those who considered man from the point of view, not of vulgar prejudice, but of science, opinions lay poles asunder. Linnæus had taken one view, Cuvier another; and, among my senior contemporaries, men like Lyell, regarded by many as revolutionaries of the deepest dye, were strongly opposed to anything which tended to break down the barrier between man and the rest of the animal world.

My own mind was by no means definitely made up about this matter when, in the year 1857, a paper was read before the Linnæan Society "On the Characters, Principles of Division and Primary Groups of the Class Mammalia," in which certain anatomical features of the brain were said to be "peculiar to the genus *Homo*," and were made the chief ground for separating that genus from all other mammals, and placing him in a division, "Archencephala," apart from, and superior to, all the rest. As these statements did not agree with the opinions I had formed, I set to work to reinvestigate the subject; and soon satisfied myself that the structures in question were not peculiar to Man, but were shared by him with all the higher and many of the lower apes. I embarked in no public discussion of these matters; but my attention being thus drawn to them, I studied the whole question of the structural relations of Man to the next lower existing forms, with much care. And, of course, I embodied my conclusions in my teaching.

Matters were at this point, when "The Origin of Species" appeared. The weighty sentence "Light will be thrown on the origin of man and his history" (1st ed. p. 488) was not only in full harmony with the conclusions at which I had arrived, respecting the structural relations of apes and men, but was strongly supported by them. And inasmuch as Development and Vertebrate Anatomy were not among Mr. Darwin's many specialities, it appeared to me that I should not be intruding on the ground he had made his own, if I discussed this part of the general question. In fact, I thought that I might probably serve the cause of evolution by doing so.

Some experience of popular lecturing had convinced me that the necessity of making things plain to uninstructed people, was one of the very best means of clearing up the obscure corners in one's own mind. So, in 1860, I took the Relation of Man to the Lower Animals, for the subject of the six lectures to working men which it was my duty to deliver. It was also in 1860, that this topic was discussed before a jury of experts, at the meeting of the British Association at Oxford; and, from that time, a sort of running fight on the same subject was carried on, until it culminated at the Cambridge meeting of the Association in 1862, by my friend Sir W. Flower's public demonstration of the existence in the apes of those cerebral characters which had been said to be peculiar to man.

"Magna est veritas et prævalebit!" Truth is great, certainly, but, considering her greatness, it is curious what a long time she is apt to take about prevailing. When, towards the end of 1862, I had finished writing "Man's Place in Nature," I could say with a good conscience, that my conclusions "had not been formed hastily or enunciated crudely." I thought I had earned the right to publish them and even fancied I might be thanked, rather than reproved, for so doing. However, in my anxiety to promulgate nothing erroneous, I asked a highly competent anatomist and very good friend of mine to look through my proofs and, if he could, point out any errors of fact. I was well pleased when he returned them without criticism on that score; but my satisfaction was speedily dashed by the very earnest warning, as to the consequences of publication, which my friend's interest in my welfare led him to give. But, as I have confessed elsewhere, when I was a young man, there was just a little—a mere *soupçon*—in my composition of that tenacity of purpose which has another name; and I felt sure that all the evil things prophesied would not be so painful to me as the giving up that which I had resolved to do, upon grounds which I conceived to be right. So the book came out; and I must do my friend the justice to say that his forecast was completely justified. The Boreas of criticism blew his hardest blasts of misrepresentation and ridicule for some years; and I was even as one of the wicked. Indeed, it surprises me, at times, to think how any one who had sunk so low could since have emerged into, at any rate, relative respectability. Personally, like the non-corvine personages in the Ingoldsby legend, I did not feel "one penny the worse." Translated into several languages, the book reached a wider public than I had ever hoped for; being largely helped, I imagine, by the Ernulphine advertisements to which I have referred. It has had the honour of being freely utilized, without acknowledgment, by writers of repute; and, finally, it achieved the fate, which is the euthanasia of a scientific work, of being inclosed among the rubble of the foundations of later knowledge and forgotten.

To my observation, human nature has not sensibly changed during the last thirty years. I doubt not that there are truths as plainly obvious and as generally denied, as those contained in "Man's Place in Nature," now awaiting enunciation. If there is a young man of the present generation, who has taken as much trouble as I did to assure himself that they are truths, let him come out with them, without troubling his head about the barking of the dogs of St. Ernulphus. "Veritas prævalebit"—some day; and, even if she does not prevail in his time, he himself will be all the better and the wiser for having tried to help her. And let him recollect that such great

curious what a long time she is apt to take about prevailing. When, towards the end of 1862, I had finished writing "Man's Place in Nature," I could say with a good conscience, that my conclusions "had not been formed hastily or enunciated crudely." I thought I had earned the right to publish them and even fancied I might be thanked, rather than reproved, for so doing. However, in my anxiety to promulgate nothing erroneous, I asked a highly competent anatomist and very good friend of mine to look through my proofs and, if he could, point out any errors of fact. I was well pleased when he returned them without criticism on that score; but my satisfaction was speedily dashed by the very earnest warning, as to the consequences of publication, which my friend's interest in my welfare led him to give. But, as I have confessed elsewhere, when I was a young man, there was just a little—a mere *soupçon*—in my composition of that tenacity of purpose which has another name; and I felt sure that all the evil things prophesied would not be so painful to me as the giving up that which I had resolved to do, upon grounds which I conceived to be right. So the book came out; and I must do my friend the justice to say that his forecast was completely justified. The Boreas of criticism blew his hardest blasts of misrepresentation and ridicule for some years; and I was even as one of the wicked. Indeed, it surprises me, at times, to think how any one who had sunk so low could since have emerged into, at any rate, relative respectability. Personally, like the non-corvine personages in the Ingoldsby legend, I did not feel "one penny the worse." Translated into several languages, the book reached a wider public than I had ever hoped for; being largely helped, I imagine, by the Ernulphine advertisements to which I have referred. It has had the honour of being freely utilized, without acknowledgment, by writers of repute; and, finally, it achieved the fate, which is the euthanasia of a scientific work, of being inclosed among the rubble of the foundations of later knowledge and forgotten.

To my observation, human nature has not sensibly changed during the last thirty years. I doubt not that there are truths as plainly obvious and as generally denied, as those contained in "Man's Place in Nature," now awaiting enunciation. If there is a young man of the present generation, who has taken as much trouble as I did to assure himself that they are truths, let him come out with them, without troubling his head about the barking of the dogs of St. Ernulphus. "Veritas prævalebit"—some day; and, even if she does not prevail in his time, he himself will be all the better and the wiser for having tried to help her. And let him recollect that such great reward is full payment for all his labour and pains.

"Man's Place in Nature," perhaps, may still be useful as an introduction to the subject; but, as any interest which attaches to it must be mainly historical, I have thought it right to leave the essays untouched. The history of the long controversy about the structure of the brain, following upon the second dissertation, in the original edition, however, is omitted. The verdict of science has long since been pronounced upon the questions at issue; and no good purpose can be served by preserving the memory of the details of the suit.

In many passages, the reader who is acquainted with the present state of science, will observe much room for addition; but, in all cases, the supplements required, are, I believe, either indifferent to the argument or would strengthen it.

CONTENTS

I
ON THE NATURAL HISTORY OF THE MAN-LIKE APES 1

II
ON THE RELATIONS OF MAN TO THE LOWER ANIMALS 77

III
ON SOME FOSSIL REMAINS OF MAN 157

IV
ON THE METHODS AND RESULTS OF ETHNOLOGY [1865] . 209

V
ON SOME FIXED POINTS IN BRITISH ETHNOLOGY [1871] . 253

VI
ON THE ARYAN QUESTION [1890] 271

*** The first three Essays were published in January, 1863, under the title of "Man's Place in Nature"; the fourth essay appeared in the *Fortnightly Review*, the fifth in the *Contemporary Review*, and they were republished in *Critiques and Addresses*. The Essay on the Aryan Question appeared in the *Nineteenth Century* for November, 1890.

MAN'S PLACE IN NATURE

ADVERTISEMENT TO THE READER

The greater part of the substance of the following Essays has already been published in the form of Oral Discourses, addressed to widely different audiences during the past three years.

Upon the subject of the second Essay, I delivered six Lectures to the Working Men in 1860, and two, to the members of the Philosophical Institution of Edinburgh in 1862. The readiness with which my audience followed my arguments, on these occasions, encourages me to hope that I have not committed the error, into which working men of science so readily fall, of obscuring my meaning by unnecessary technicalities: while, the length of the period during which the subject, under its various aspects has been present to my mind, may suffice to satisfy the Reader that, my conclusions, be they right or be they wrong, have not been formed hastily or enunciated crudely.

T. H. H.

LONDON: *January*, 1863.

I

ON THE NATURAL HISTORY OF THE MAN-LIKE APES

ANCIENT traditions, when tested by the severe processes of modern investigation, commonly enough fade away into mere dreams: but it is singular how often the dream turns out to have been a half-waking one, presaging a reality. Ovid foreshadowed the discoveries of the geologist: the Atlantis was an imagination, but Columbus found a western world : and though the quaint forms of Centaurs and Satyrs have an existence only in the realms of art, creatures approaching man more nearly than they in essential structure, and yet as thoroughly brutal as the goat's or horse's half of the mythical compound, are now not only known, but notorious.

I have not met with any notice of one of these MAN-LIKE APES of earlier date than that contained in Pigafetta's "Description of the kingdom of Congo,"[1] drawn up from the notes of a Portuguese sailor, Eduardo Lopez, and published in 1598. The tenth chapter of this work is entitled "De Animalibus quæ in hac provincia

Fig. 1.—Simiæ magnatum deliciæ.—De Bry, 1598.

reperiuntur," and contains a brief passage to the effect that " in the Songan country, on the banks of the Zaire, there are multitudes of apes, which

[1] REGNUM CONGO: hoc est VERA DESCRIPTIO REGNI AFRICANI QUOD TAM AB INCOLIS QUAM LUSITANIS CONGUS APPELLATUR, per Philippum Pigafettam, olim ex Edoardo Lopez acroamatis lingua Italica excerpta, num Latio sermone donata ab August. Cassiod. Reinio. Iconibus et imaginibus rerum memorabilium quasi vivis, opera et industria Joan. Theodori et Joan. Israelis de Bry, fratrum exornata. Francofurti, MDXCVIII.

afford great delight to the nobles by imitating human gestures." As this might apply to almost any kind of apes, I should have thought little of it, had not the brothers De Bry, whose engravings illustrate the work, thought fit, in their eleventh "Argumentum," to figure two of these "Simiæ magnatum deliciæ." So much of the plate as contains these apes is faithfully copied in the woodcut (Fig. 1), and it will be observed that they are tail-less, long-armed, and large-eared; and about the size of Chimpanzees. It may be that these apes are as much figments of the imagination of the ingenious brothers as the winged, two-legged, crocodile-headed dragon which adorns the same plate; or, on the other hand, it may be that the artists have constructed their drawings from some essentially faithful description of a Gorilla or a Chimpanzee. And, in either case, though these figures are worth a passing notice, the oldest trustworthy and definite accounts of any animal of this kind date from the 17th century, and are due to an Englishman.

The first edition of that most amusing old book, "Purchas his Pilgrimage," was published in 1613, and therein are to be found many references to the statements of one whom Purchas terms "Andrew Battell (my neere neighbour, dwelling at Leigh in Essex) who served under Manuel Silvera Perera, Governor under the King of Spaine, at his city of Saint Paul, and with him went farre into the countrey of Angola"; and again, "my friend, Andrew Battle, who lived in the kingdom of Congo many yeares," and who, "upon some quarell betwixt the Portugals (among whom he was a sergeant of a band) and him, lived eight or nine moneths in the woodes." From this weather-beaten old soldier, Purchas was amazed to hear "of a kinde of Great Apes, if they might so bee termed, of the height of a man, but twice as bigge in feature of their limmes, with strength proportionable, hairie all over. otherwise altogether like men and women in their whole bodily shape.[1] They lived on such wilde fruits as the trees and woods yielded, and in the night time lodged on the trees."

This extract is, however, less detailed and clear in its statements than a passage in the third chapter of the second part of another work—"Purchas his Pilgrimes," published in 1625, by the same author—which has been often, though hardly ever quite rightly, cited. The chapter is entitled, "The strange adventures of Andrew Battell, of Leigh in Essex, sent by the Portugals prisoner to Angola, who lived there and in the adioining regions neere eighteene yeeres." And the sixth section of this chapter is headed—"Of the Provinces of Bongo, Calongo, Mayombe, Manikesocke, Motimbas: of the Ape Monster Pongo,

---

[1] "Except this that their legges had no calves."—[Ed. 1626.] And in a marginal note, "These great apes are called Pongo's."

their hunting: Idolatries; and divers other observations."

"This province (Calongo) toward the east bordereth upon Bongo, and toward the north upon Mayombe, which is nineteen leagues from Longo along the coast.

"This province of Mayombe is all woods and groves, so overgrowne that a man may travaile twentie days in the shadow without any sunne or heat. Here is no kind of corne nor graine, so that the people liveth onely upon plantanes and roots of sundrie sorts, very good; and nuts; nor any kinde of tame cattell, nor hens.

"But they have great store of elephants' flesh, which they greatly esteeme, and many kinds of wild beasts; and great store of fish. Here is a great sandy bay, two leagues to the northward of Cape Negro,[1] which is the port of Mayombe. Sometimes the Portugals lade logwood in this bay. Here is a great river, called Banna: in the winter it hath no barre, because the generall winds cause a great sea. But when the sunne hath his south declination, then a boat may goe in; for then it is smooth because of the raine. This river is very great, and hath many ilands and people dwelling in them. The woods are so covered with baboones, monkies, apes and parrots, that it will feare any man to travaile in them alone. Here are also two kinds of monsters, which are common in these woods, and very dangerous.

"The greatest of these two monsters is called Pongo in their language, and the lesser is called Engeco. This Pongo is in all proportion like a man; but that he is more like a giant in stature than a man; for he is very tall, and hath a man's face, hollow-eyed, with long haire upon his browes. His face and eares are without haire, and his hands also. His bodie is full of haire, but not very thicke; and it is of a dunnish colour.

"He differeth not from a man but in his legs; for they have

---

[1] *Purchas' note.*—Cape Negro is in 16 degrees south of the line.

no calfe. Hee goeth alwaies upon his legs, and carrieth his hands clasped in the nape of his necke when he goeth upon the ground. They sleepe in the trees, and build shelters for the raine. They feed upon fruit that they find in the woods, and upon nuts, for they eate no kind of flesh. They cannot speake, and have no understanding more than a beast. The people of the countrie, when they travaile in the woods make fires where they sleepe in the night; and in the morning when they are gone, the Pongoes will come and sit about the fire till it goeth out; for they have no understanding to lay the wood together. They goe many together and kill many negroes that travaile in the woods. Many times they fall upon the elephants which come to feed where they be, and so beate them with their clubbed fists, and pieces of wood, that they will runne roaring away from them. Those Pongoes are never taken alive because they are so strong, that ten men cannot hold one of them; but yet they take many of their young ones with poisoned arrowes.

"The young Pongo hangeth on his mother's belly with his hands fast clasped about her, so that when the countrie people kill any of the females they take the young one, which hangeth fast upon his mother.

"When they die among themselves, they cover the dead with great heaps of boughs and wood, which is commonly found in the forest."[1]

It does not appear difficult to identify the exact region of which Battell speaks. Longo is

[1] *Purchas' marginal note*, p. 982:—"The Pongo a giant ape. He told me in conference with him, that one of these Pongoes tooke a negro boy of his which lived a moneth with them. For they hurt not those which they surprise at unawares, except they look on them; which he avoyded. He said their highth was like a man's, but their bignesse twice as great. I saw the negro boy. What the other monster should be he hath forgotten to relate; and these papers came to my hand since his death, which, otherwise, in my often conferences, I might have learned. Perhaps he meaneth the Pigmy Pongo killers mentioned."

doubtless the name of the place usually spelled Loango on our maps. Mayombe still lies some nineteen leagues northward from Loango, along the coast; and Cilongo or Kilonga, Manikesocke, and Motimbas are yet registered by geographers. The Cape Negro of Battell, however, cannot be the modern Cape Negro in 16° S., since Loango itself is in 4° S. latitude. On the other hand, the "great river called Banna" corresponds very well with the "Camma" and "Fernand Vas," of modern geographers, which form a great delta on this part of the African coast.

Now this "Camma" country is situated about a degree and a-half south of the Equator, while a few miles to the north of the line lies the Gaboon, and a degree or so north of that, the Money River—both well known to modern naturalists as localities where the largest of man-like Apes has been obtained. Moreover, at the present day, the word Engeco, or N'schego, is applied by the natives of these regions to the smaller of the two great Apes which inhabit them; so that there can be no rational doubt that Andrew Battell spoke of that which he knew of his own knowledge, or, at any rate, by immediate report from the natives of Western Africa. The "Engeco," however, is that "other monster" whose nature Battell "forgot to relate," while the name "Pongo"—applied to the animal whose characters and habits are so fully and carefully described—seems to have died out, at least in its primitive form and signification. Indeed, there is evidence that not only in Battell's time, but up to a very recent date, it was used in a totally different sense from that in which he employs it.

For example, the second chapter of Purchas' work, which I have just quoted, contains "A Description and Historicall Declaration of the Golden Kingdom of Guinea, &c. &c. Translated from the Dutch, and compared also with the Latin," wherein it is stated (p. 986) that—

"The River Gaboon lyeth about fifteen miles northward from Rio de Angra, and eight miles northward from Cape de Lope Gonsalvez (Cape Lopez), and is right under the Equinoctial line, about fifteene miles from St. Thomas, and is a great land, well and easily to be knowne. At the mouth of the river there lieth a sand, three or foure fathoms deepe, whereon it beateth mightily with the streame which runneth out of the river into the sea. This river, in the mouth thereof, is at least four miles broad; but when you are about the Iland called *Pongo*, it is not above two miles broad. . . . On both sides the river there standeth many trees. . . . . . . The Iland called *Pongo*, which hath a monstrous high hill."

The French naval officers, whose letters are appended to the late M. Isidore Geoff. Saint Hilaire's excellent essay on the Gorilla,[1] note in similar terms the width of the Gaboon, the trees that line its banks down to the water's edge, and the strong current that sets out of it. They describe two islands in its estuary;—one low,

[1] *Archives du Muscum*, Tome X.

called Perroquet; the other high, presenting three conical hills, called Coniquet; and one of them, M. Franquet, expressly states that, formerly, the Chief of Coniquet was called *Meni-Pongo*, meaning thereby Lord of *Pongo;* and that the *N'Pongues* (as, in agreement with Dr. Savage, he affirms the natives call themselves) term the estuary of the Gaboon itself *N'Pongo*.

It is so easy, in dealing with savages, to misunderstand their applications of words to things, that one is at first inclined to suspect Battell of having confounded the name of this region, where his "greater monster" still abounds, with the name of the animal itself. But he is so right about other matters (including the name of the "lesser monster") that one is loth to suspect the old traveller of error; and, on the other hand, we shall find that a voyager of a hundred years' later date speaks of the name "Boggoe," as applied to a great Ape, by the inhabitants of quite another part of Africa—Sierra Leone.

But I must leave this question to be settled by philologers and travellers; and I should hardly have dwelt so long upon it except for the curious part played by this word '*Pongo*' in the later history of the man-like Apes.

The generation which succeeded Battell saw the first of the man-like Apes which was ever brought to Europe, or, at any rate, whose visit found a historian. In the third book of Tulpius'

"Observationes Medicæ," published in 1641, the 56th chapter or section is devoted to what he calls *Satyrus indicus*, "called by the Indians Orang-autang or Man-of-the-Woods, and by the Africans Quoias Morrou." He gives a very good

Fig. 2.—The Orang of Tulpius, 1641.

figure, evidently from the life, of the specimen of this animal, "nostra memoria ex Angolâ delatum," presented to Frederick Henry Prince of Orange. Tulpius says it was as big as a child of three years old, and as stout as one of six years: and that its back was covered with black hair. It is plainly a young Chimpanzee.

In the meanwhile, the existence of other, Asiatic, man-like Apes became known, but at first in a very mythical fashion. Thus Bontius (1658) gives an altogether fabulous and ridiculous account and figure of an animal which he calls "Orang-outang"; and though he says "vidi Ego cujus effigiem hic exhibeo," the said effigies (see Fig. 6 for Hoppius' copy of it) is nothing but a very hairy woman of rather comely aspect, and with proportions and feet wholly human. The judicious English anatomist, Tyson, was justified in saying of this description by Bontius, "I confess I do mistrust the whole representation."

It is to the last-mentioned writer, and his coadjutor Cowper, that we owe the first account of a man-like ape which has any pretensions to scientific accuracy and completeness. The treatise entitled, "*Orang-outang, sive Homo Sylvestris;* or the Anatomy of a Pygmie compared with that of a *Monkey,* an *Ape,* and a *Man,*" published by the Royal Society in 1699, is, indeed, a work of remarkable merit, and has, in some respects, served as a model to subsequent inquirers. This "Pygmie," Tyson tells us "was brought from Angola, in Africa; but was first taken a great deal higher up the country"; its hair "was of a coal-black colour and strait," and "when it went as a quadruped on all four, 'twas awkwardly; not placing the palm of the hand flat to the ground, but it walk'd upon its knuckles, as I observed it to do when weak and had not strength enough to support its body,"—" From

FIG. 3.—The "Pygmie" reduced from Tyson's figure 1, 1699.

the top of the head to the heel of the foot, in a strait line, it measured twenty-six inches."

These characters, even without Tyson's good figure (Figs. 3 and 4), would have been sufficient to prove his "Pygmie" to be a young Chimpanzee. But the opportunity of examining the skeleton of the very animal Tyson anatomised having most unexpectedly presented itself to me, I am able to

Fig. 4.—The "Pygmie" reduced from Tyson's figure 2, 1699.

bear independent testimony to its being a veritable *Troglodytes niger*,[1] though still very young.

[1] I am indebted to Dr. Wright, of Cheltenham, whose paleontological labours are so well known, for bringing this

Although fully appreciating the resemblances between his Pygmie and Man, Tyson by no means overlooked the differences between the two, and he concludes his memoir by summing up first, the points in which "the Ourang-outang or Pygmie more resembled a Man than Apes and Monkeys do," under forty-seven distinct heads; and then giving, in thirty-four similar brief paragraphs, the respects in which "the Ourang-outang or Pygmie differ'd from a man and resembled more the Ape and Monkey kind."

After a careful survey of the literature of the subject extant in his time, our author arrives at the conclusion that his "Pygmie" is identical neither with the Orangs of Tulpius and Bontius, nor with the Quoias Morrou of Dapper (or rather of Tulpius), the Barris of d'Arcos, nor with the Pongo of Battell; but that it is a species of ape probably identical with the Pygmies of the Ancients, and, says Tyson, though it "does so much resemble *a Man* in many of its parts, more than any of the ape kind, or any other *animal* in the world, that I know of: yet by no means do I look upon it as the product of a *mixt* generation—

---

interesting relic to my knowledge. Tyson's granddaughter, it appears, married Dr. Allardyce, a physician of repute in Cheltenham, and brought, as part of her dowry, the skeleton of the "Pygmie." Dr. Allardyce presented it to the Cheltenham Museum, and, through the good offices of my friend Dr. Wright, the authorities of the Museum have permitted me to borrow, what is, perhaps, its most remarkable ornament.

'tis a *Brute-Animal sui generis*, and a particular species of Ape."

The name of "Chimpanzee," by which one of the African Apes is now so well known, appears to have come into use in the first half of the eighteenth century, but the only important addition made, in that period, to our acquaintance with the man-like apes of Africa is contained in " A New Voyage to Guinea," by William Smith, which bears the date 1744.

In describing the animals of Sierra Leone, p. 51, this writer says :—

"I shall next describe a strange sort of animal, called by the white men in this country Mandrill,[1] but why it is so called I know not, nor did I ever hear the name before, neither can those who call them so tell, except it be for their near resemblance of a human creature, though nothing at all like an Ape. Their bodies, when full grown, are as big in circumference as a middle-sized man's—their legs much shorter, and their feet larger; their arms and hands in proportion. The head is monstrously big, and the face broad and flat, without any other hair but the eyebrows; the nose very small, the mouth wide,

---

[1] "Mandrill" seems to signify a "man-like ape," the word "Drill" or "Dril" having been anciently employed in England to denote an Ape or Baboon. Thus in the fifth edition of Blount's *"Glossographia*, or a Dictionary interpreting the hard words of whatsoever language now used in our refined English tongue . . . very useful for all such as desire to understand what they read," published in 1681, I find, "Dril—a stonecutter's tool wherewith he bores little holes in marble, &c. Also a large overgrown Ape and Baboon, so called." "Drill" is used in the same sense in Charleton's *Onomasticon Zoicon*, 1668. The singular etymology of the word given by Buffon seems hardly a probable one.

and the lips thin. The face, which is covered by a white skin, is monstrously ugly, being all over wrinkled as with old age; the teeth broad and yellow; the hands have no more hair than the face, but the same white skin, though all the rest of the body is covered with long black hair, like a bear. They never go upon all-fours, like apes; but cry, when vexed or teased, just like children. . . . . . .

FIG. 5.—Facsimile of William Smith's figure of the "Mandrill," 1744.

'When I was at Sherbro, one Mr. Cummerbus, whom I shall have occasion hereafter to mention, made me a present of one of these strange animals, which are called by the natives Boggoe: it was a she-cub, of six months' age, but even then larger than a Baboon. I gave it in charge to one of the slaves, who knew how to feed and nurse it, being a very tender sort of animal; but whenever I went off the deck the sailors began to teaze it—some loved to see its tears and hear it cry; others hated its snotty nose; one who hurt it, being checked by the negro that took care of it, told the slave he was very fond of his country-woman, and asked him if he should not like her for a wife? To which the slave very readily replied, 'No, this no my wife; this a white woman—this fit wife for you.' This unlucky wit of the negro's, I fancy, hastened its death, for next morning it was found dead under the windlass."

William Smith's "Mandrill," or "Boggoe," as his description and figure testify, was, without doubt, a Chimpanzee.

Linnæus knew nothing, of his own observation, of the man-like Apes of either Africa or Asia, but a dissertation by his pupil Hoppius in the "Amœnitates Academicæ" (VI. "Anthropomorpha") may be regarded as embodying his views respecting these animals.

The dissertation is illustrated by a plate, of which the accompanying woodcut, Fig. 6, is a reduced copy. The figures are entitled (from left to right 1. *Troglodyta Bontii*; 2. *Lucifer Aldrovandi*; 3. *Satyrus Tulpii*; 4. *Pygmæus Edwardi*. The first is a bad copy of Bontius' fictitious "Ourang-outang," in whose existence, however, Linnæus appears to have fully believed; for in the standard edition of the "Systema Naturæ," it is enumerated as a second species of Homo; "H. nocturnus." *Lucifer Aldrovandi* is a copy of a figure in Aldrovandus, "De Quadrupedibus digitatis viviparis," Lib. 2, p. 249 (1645) entitled "Cercopithecus formæ raræ *Barbilius* vocatus et originem a china ducebat." Hoppius is of opinion that this may be one of that cat-tailed people, of whom Nicolaus Köping affirms that they eat a boat's crew, "gubernator navis" and all! In the "Systema Naturæ" Linnæus calls it in a note, *Homo caudatus*, and seems inclined to regard it as a third species of man. According to Temminck, *Satyrus Tulpii* is a copy of the figure

Fig. 6.—The Anthropomorpha of Linnæus.

of a Chimpanzee published by Scotin in 1738, which I have not seen. It is the *Satyrus indicus* of the "Systema Naturæ," and is regarded by Linnæus as possibly a distinct species from *Satyrus sylvestris*. The last, named *Pygmæus Edwardi*, is copied from the figure of a young "Man of the Woods," or true Orang-Utan, given in Edwards' "Gleanings of Natural History" (1758).

Buffon was more fortunate than his great rival. Not only had he the rare opportunity of examining a young Chimpanzee in the living state, but he became possessed of an adult Asiatic man-like Ape—the first and the last adult specimen of any of these animals brought to Europe for many years. With the valuable assistance of Daubenton, Buffon gave an excellent description of this creature, which, from its singular proportions, he termed the long-armed Ape, or Gibbon. It is the modern *Hylobates lar*.

Thus when, in 1766, Buffon wrote the fourteenth volume of his great work, he was personally familiar with the young of one kind of African man-like Ape, and with the adult of an Asiatic species—while the Orang-Utan and the Mandrill of Smith were known to him by report. Furthermore, the Abbé Prevost had translated a good deal of Purchas' "Pilgrims" into French, in his "Histoire générale des Voyages" (1748), and there Buffon found a version of Andrew Battell's account of the Pongo and the Engeco. All these data Buffon attempts to weld together into harmony in this chapter entitled "Les Orang-outangs ou le Pongo et le Jocko." To this title the following note is appended :—

"Orang-outang nom de cet animal aux Indes orientales: Pongo nom de cet animal à Lowando Province de Congo.

"Jocko, Enjocko, nom de cet animal à Congo que nous avons adopté. *En* est l'article que nous avons retranché."

Thus it was that Andrew Battell's "Engeco" became metamorphosed into "Jocko," and, in the latter shape, was spread all over the world, in consequence of the extensive popularity of Buffon's works. The Abbé Prevost and Buffon between them however, did a good deal more disfigurement to Battell's sober account than "cutting off an article." Thus Battell's statement that the Pongos "cannot speake, and have no understanding more than a beast," is rendered by Buffon "qu'il ne peut parler *quoiqu'il ait plus d'entendement que les autres animaux;*" and again, Purchas' affirmation, "He told me in conference with him, that one of these Pongos tooke a negro boy of his which lived a moneth with them," stands in the French version, "un pongo lui enleva un petit negre qui passa un *an* entier dans la société de ces animaux."

After quoting the account of the great Pongo, Buffon justly remarks, that all the "Jockos" and "Orangs" hitherto brought to Europe were young; and he suggests that, in their adult condition, they might be as big as the Pongo or "great Orang;" so that, provisionally, he regarded the Jockos, Orangs, and Pongos as all of one species. And perhaps this was as much as the state of knowledge at the time warranted. But how it came about that Buffon failed to perceive the similarity of Smith's "Mandrill" to his own "Jocko," and confounded the former with so totally different a creature as the blue-faced Baboon, is not so easily intelligible.

Twenty years later Buffon changed his opinion,[1] and expressed his belief that the Orangs constituted a genus with two species,—a large one, the Pongo of Battell, and a small one, the Jocko: that the small one (Jocko) is the East Indian Orang; and that the young animals from Africa, observed by himself and Tulpius, are simply young Pongos.

In the meanwhile, the Dutch naturalist, Vosmaer, gave, in 1778, a very good account and figure of a young Orang, brought alive to Holland, and his countryman, the famous anatomist, Peter Camper, published (1779) an essay on the Orang-Utan of similar value to that of Tyson on the Chimpanzee. He dissected several females and a male, all of which, from the state of their skeleton and their dentition, he justly supposes to have been young. However, judging by the analogy of man, he concludes that they could not have exceeded four feet in height in the adult condition. Furthermore, he is very clear as to the specific distinctness of the true East Indian Orang.

"The Orang," says he, "differs not only from the Pigmy of Tyson and from the Orang of Tulpius by its peculiar colour and its long toes, but also by its whole external form. Its arms, its

[1] *Histoire Naturelle*, Suppl. Tome 7ème, 1789.

hands, and its feet are longer, while the thumbs, on the contrary, are much shorter, and the great toes much smaller in proportion."[1] And again, "The true Orang, that is to say, that of Asia, that of Borneo, is consequently not the Pithecus, or tail-less Ape, which the Greeks, and especially Galen, have described. It is neither the Pongo nor the Jocko, nor the Orang of Tulpius, nor the Pigmy of Tyson,—*it is an animal of a peculiar species*, as I shall prove in the clearest manner by the organs of voice and the skeleton in the following chapters" (*l. c.* p. 64).

A few years later, M. Radermacher, who held a high office in the Government of the Dutch dominions in India, and was an active member of the Batavian Society of Arts and Sciences, published, in the second part of the Transactions of that Society,[2] a Description of the Island of Borneo, which was written between the years 1779 and 1781, and, among much other interesting matter, contains some notes upon the Orang. The small sort of Orang-Utan, viz. that of Vosmaer and of Edwards, he says, is found only in Borneo, and chiefly about Banjermassing, Mampauwa, and Landak. Of these he had seen some fifty during his residence in the Indies; but none exceeded 2½ feet in length. The larger sort,

---

[1] Camper, *Œuvres*, i., p. 56.
[2] *Verhandelingen van het Bataviaasch Genootschap.* Tweede Deel. Derde Druk. 1826.

often regarded as a chimæra, continues Radermacher, would perhaps long have remained so, had it not been for the exertions of the Resident at Rembang, M. Palm, who, on returning from Landak towards Pontiana, shot one, and forwarded it to Batavia in spirit, for transmission to Europe.

Palm's letter describing the capture runs thus:—" Herewith I send your Excellency, contrary to all expectation (since long ago I offered more than a hundred ducats to the natives for an Orang-Utan of four or five feet high) an Orang which I heard of this morning about eight o'clock. For a long time we did our best to take the frightful beast alive in the dense forest about half way to Landak. We forgot even to eat, so anxious were we not to let him escape; but it was necessary to take care that he did not revenge himself, as he kept continually breaking off heavy pieces of wood and green branches, and dashing them at us. This game lasted till four o'clock in the afternoon, when we determined to shoot him; in which I succeeded very well, and indeed better than I ever shot from a boat before; for the bullet went just into the side of his chest, so that he was not much damaged. We got him into the prow still living, and bound him fast, and next morning he died of his wounds. All Pontiana came on board to see him when we arrived." Palm gives his height from the head to the heel as 49 inches.

A very intelligent German officer, Baron Von Wurmb, who at this time held a post in the Dutch East India service, and was Secretary of the Batavian Society, studied this animal, and his careful description of it, entitled "Beschrijving van der Groote Borneosche Orang-outang of de Oost-Indische Pongo," is contained in the same volume of the Batavian Society's Transactions. After Von Wurmb had drawn up his description he states, in a letter dated Batavia, Feb. 18, 1781,[1] that the specimen was sent to Europe in brandy to be placed in the collection of the Prince of Orange; "unfortunately," he continues, "we hear that the ship has been wrecked." Von Wurmb died in the course of the year 1781, the letter in which this passage occurs being the last he wrote: but in his posthumous papers, published in the fourth part of the Transactions of the Batavian Society, there is a brief description, with measurements, of a female Pongo four feet high.

Did either of these original specimens, on which Von Wurmb's descriptions are based, ever reach Europe? It is commonly supposed that they did; but I doubt the fact. For, appended to the memoir "De l'Ourang-outang," in the collected edition of Camper's works, tome i., pp. 64–66, is a note by Camper himself,

[1] "Briefe des Herrn v. Wurmb und des H. Baron von Wollzogen. Gotha, 1794."

referring to Von Wurmb's papers, and continuing thus:—"Heretofore, this kind of ape had never been known in Europe. Radermacher has had the kindness to send me the skull of one of these animals, which measured fifty-three inches, or four feet five inches, in height. I have sent some sketches of it to M. Soemmering at

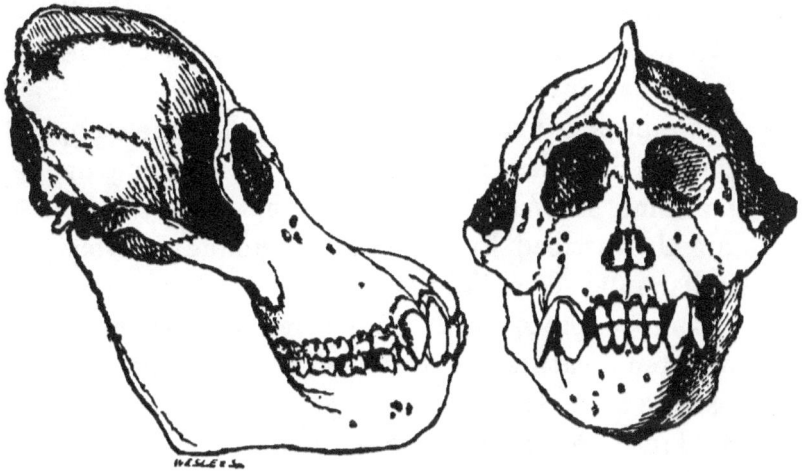

Fig. 7.—The Pongo Skull, sent by Radermacher to Camper, after Camper's original sketches, as reproduced by Lucæ.

Mayence, which are better calculated, however, to give an idea of the form than of the real size of the parts."

These sketches have been reproduced by Fischer and by Lucæ, and bear date 1783, Soemmering having received them in 1784. Had either of Von Wurmb's specimens reached

Holland, they would hardly have been unknown at this time to Camper, who, however, goes on to say:—" It appears that since this, some more of these monsters have been captured, for an entire skeleton, very badly set up, which had been sent to the Museum of the Prince of Orange, and which I saw only on the 27th of June, 1784, was more than four feet high. I examined this skeleton again on the 19th December, 1785, after it had been excellently put to rights by the ingenious Onymus."

It appears evident, then, that this skeleton, which is doubtless that which has always gone by the name of Wurmb's Pongo, is not that of the animal described by him, though unquestionably similar in all essential points.

Camper proceeds to note some of the most important features of this skeleton; promises to describe it in detail by-and-bye; and is evidently in doubt as to the relation of this great "Pongo" to his "petit Orang."

The promised further investigations were never carried out; and so it happened that the Pongo of Von Wurmb took its place by the side of the Chimpanzee, Gibbon, and Orang as a fourth and colossal species of man-like Ape. And indeed nothing could look much less like the Chimpanzees or the Orangs, then known, than the Pongo; for all the specimens of Chimpanzee and Orang which had been observed were small of stature, singularly human in aspect, gentle and docile; while Wurmb's Pongo was a monster almost twice their size, of vast strength and fierceness, and very brutal in expression; its great projecting muzzle, armed with strong teeth, being further disfigured by the outgrowth of the cheeks into fleshy lobes.

Eventually, in accordance with the usual marauding habits of the Revolutionary armies, the "Pongo" skeleton was carried away from Holland into France, and notices of it, expressly intended to demonstrate its entire distinctness from the Orang and its affinity with the baboons, were given, in 1798, by Geoffroy St. Hilaire and Cuvier.

Even in Cuvier's "Tableau Élémentaire," and in the first edition of his great work, the "Regne Animal," the "Pongo" is classed as a species of Baboon. However, so early as 1818, it appears that Cuvier saw reason to alter this opinion, and to adopt the view suggested several years before by Blumenbach,[1] and after him by Tilesius, that the Bornean Pongo is simply an adult Orang. In 1824, Rudolphi demonstrated, by the condition of the dentition, more fully and completely than had been done by his predecessors, that the Orangs described up to that time were all young animals, and that the skull and teeth of the adult

[1] See Blumenbach *Abbildungen Naturhistorichen Gegenstande*, No. 12, 1810; and Tilesius, *Naturhistoriche Früchte der ersten Kaiserlich-Russischen Erdumsegelung*, p. 115, 1813.

would probably be such as those seen in the Pongo of Wurmb. In the second edition of the "Regne Animal" (1829), Cuvier infers, from the "proportions of all the parts" and "the arrangements of the foramina and sutures of the head," that the Pongo is the adult of the Orang-Utan, "at least of a very closely allied species," and this conclusion was eventually placed beyond all doubt by Professor Owen's Memoir published in the "Zoological Transactions" for 1835, and by Temminck in his "Monographies de Mammalogie." Temminck's memoir is remarkable for the completeness of the evidence which it affords as to the modification which the form of the Orang undergoes according to age and sex. Tiedemann first published an account of the brain of the young Orang, while Sandifort, Müller and Schlegel, described the muscles and the viscera of the adult, and gave the earliest detailed and trustworthy history of the habits of the great Indian Ape in a state of nature; and as important additions have been made by later observers, we are at this moment better acquainted with the adult of the Orang-Utan, than with that of any of the other greater man-like Apes.

It is certainly the Pongo of Wurmb;[1] and it is as certainly not the Pongo of Battell, seeing that

[1] Speaking broadly and without prejudice to the question, whether there be more than one species of Orang

the Orang-Utan is entirely confined to the great Asiatic islands of Borneo and Sumatra.

And while the progress of discovery thus cleared up the history of the Orang, it also became established that the only other man-like Apes in the eastern world were the various species of Gibbon—Apes of smaller stature, and therefore attracting less attention than the Orangs, though they are spread over a much wider range of country, and are hence more accessible to observation.

Although the geographical area inhabited by the "Pongo" and "Engeco" of Battell is so much nearer to Europe than that in which the Orang and Gibbon are found, our acquaintance with the African Apes has been of slower growth; indeed, it is only within the last few years that the truthful story of the old English adventurer has been rendered fully intelligible. It was not until 1835 that the skeleton of the adult Chimpanzee became known, by the publication of Professor Owen's above-mentioned very excellent memoir "On the Osteology of the Chimpanzee and Orang," in the Zoological Transactions—a memoir which, by the accuracy of its descriptions, the carefulness of its comparisons, and the excellence of its figures, made an epoch in the history of our knowledge of the bony framework, not only of the Chimpanzee, but of all the anthropoid Apes.

By the investigations herein detailed, it became

evident that the old Chimpanzee acquired a size and aspect as different from those of the young known to Tyson, to Buffon, and to Traill, as those of the old Orang from the young Orang; and the subsequent very important researches of Messrs. Savage and Wyman, the American missionary and anatomist, have not only confirmed this conclusion, but have added many new details.[1]

One of the most interesting among the many valuable discoveries made by Dr. Thomas Savage is the fact, that the natives in the Gaboon country at the present day, apply to the Chimpanzee a name—"Enché-eko"—which is obviously identical with the "Engeko" of Battell; a discovery which has been confirmed by all later inquirers. Battell's "lesser monster" being thus proved to be a veritable existence, of course a strong presumption arose that his "greater monster," the "Pongo," would sooner or later be discovered. And, indeed, a modern traveller, Bowdich, had, in 1819, found strong evidence, among the natives, of the existence of a second great Ape, called the "Ingena," "five feet high, and four across the shoulders," the builder of a rude house, on the outside of which it slept.

---

[1] See "Observations on the external characters and habits of the Troglodytes niger, by Thomas N. Savage, M.D., and on its organization, by Jeffries Wyman, M.D.," *Boston Journal of Natural History*, vol. iv. 1843-4; and "External characters, habits, and osteology of Troglodytes Gorilla," by the same authors, ibid. vol. v. 1847.

In 1847, Dr. Savage had the good fortune to make another and most important addition to our knowledge of the man-like Apes; for, being unexpectedly detained at the Gaboon river, he saw in the house of the Rev. Mr. Wilson, a missionary resident there, "a skull represented by the natives to be a monkey-like animal, remarkable for its size, ferocity, and habits." From the contour of the skull, and the information derived from several intelligent natives, "I was induced," says Dr. Savage (using the term Orang in its old general sense) "to believe that it belonged to a new species of Orang. I expressed this opinion to Mr. Wilson, with a desire for further investigation; and, if possible, to decide the point by the inspection of a specimen alive or dead." The result of the combined exertions of Messrs. Savage and Wilson was not only the obtaining of a very full account of the habits of this new creature, but a still more important service to science, the enabling the excellent American anatomist already mentioned, Professor Wyman, to describe, from ample materials, the distinctive osteological characters of the new form. This animal was called by the natives of the Gaboon "Engé-ena," a name obviously identical with the "Ingena" of Bowdich; and Dr. Savage arrived at the conviction that this last discovered of all the great Apes was the long-sought "Pongo" of Battell.

The justice of this conclusion, indeed, is beyond

doubt—for not only does the "Engé-ena" agree with Battell's "greater monster" in its hollow eyes, its great stature, and its dun or iron-grey colour, but the only other man-like Ape which inhabits these latitudes—the Chimpanzee—is at once identified, by its smaller size, as the "lesser monster," and is excluded from any possibility of being the "Pongo," by the fact that it is black and not dun, to say nothing of the important circumstance already mentioned that it still retains the name of "Engeko," or "Enché-eko," by which Battell knew it.

In seeking for a specific name for the "Engeena," however, Dr. Savage wisely avoided the much misused "Pongo"; but finding in the ancient Periplus of Hanno the word "Gorilla" applied to certain hairy savage people, discovered by the Carthaginian voyager in an island on the African coast, he attached the specific name "*Gorilla*" to his new ape, whence arises its present well-known appellation. But Dr. Savage, more cautious than some of his successors, by no means identifies his ape with Hanno's "wild men." He merely says that the latter were "probably one of the species of the Orang;" and I quite agree with M. Brullé, that there is no ground for identifying the modern "Gorilla" with that of the Carthaginian admiral.

Since the memoir of Savage and Wyman was published, the skeleton of the Gorilla has been investigated by Professor Owen and by the late Professor Duvernoy, of the Jardin des Plantes, the latter having further supplied a valuable account of the muscular system and of many of the other soft parts; while African missionaries and travellers have confirmed and expanded the account originally given of the habits of this great man-like Ape, which has had the singular fortune of being the first to be made known to the general world and the last to be scientifically investigated.

Two centuries and a half have passed away since Battell told his stories about the "greater" and the "lesser monsters" to Purchas, and it has taken nearly that time to arrive at the clear result that there are four distinct kinds of Anthropoids—in Eastern Asia, the Gibbons and the Orangs; in Western Africa, the Chimpanzees and the Gorilla.

The man-like Apes, the history of the discovery of which has just been detailed, have certain characters of structure and of distribution in common. Thus they all have the same number of teeth as man—possessing four incisors, two canines, four false molars, and six true molars in each jaw, or 32 teeth in all, in the adult condition; while the milk dentition consists of 20 teeth—or four incisors, two canines, and four molars in each jaw. They are what are called catarrhine Apes—that is, their nostrils have a narrow partition and look downwards; and, furthermore, their arms are always longer than their legs, the difference being sometimes greater and sometimes less; so that if the four were arranged in the order of the length of their arms in proportion to that of their legs, we should have this series—Orang ($1\frac{2}{8}$—1), Gibbon ($1\frac{1}{4}$—1), Gorilla ($1\frac{1}{8}$—1), Chimpanzee ($1\frac{1}{16}$—1). In all, the fore limbs are terminated by hands, provided with longer or shorter thumbs; while the great toe of the foot, always smaller than in Man, is far more movable than in him and can be opposed, like a thumb, to the rest of the foot. None of these apes have tails, and none of them possess the cheek-pouches common among monkeys. Finally, they are all inhabitants of the old world.

The Gibbons are the smallest, slenderest, and longest-limbed of the man-like apes: their arms are longer in proportion to their bodies than those of any of the other man-like Apes, so that they can touch the ground when erect; their hands are longer than their feet, and they are the only Anthropoids which possess callosities like the lower monkeys. They are variously coloured. The Orangs have arms which reach to the ankles in the erect position of the animal; their thumbs and great toes are very short, and their feet are longer than their hands. They are covered with reddish brown hair, and the sides of the face, in adult males, are commonly produced into two crescentic, flexible excrescences, like fatty tumours. The Chimpanzees have arms which reach below the knees; they have large thumbs and great toes; their hands are longer than their feet; and their hair is black, while the skin of the face is pale. The Gorilla, lastly, has arms which reach to the middle of the leg, large thumbs and great toes, feet longer than the hands, a black face, and dark-grey or dun hair.

For the purpose which I have at present in view, it is unnecessary that I should enter into any further minutiæ respecting the distinctive characters of the genera and species into which these man-like Apes are divided by naturalists. Suffice it to say, that the Orangs and the Gibbons constitute the distinct genera, *Simia* and *Hylobates;* while the Chimpanzees and Gorillas are by some regarded simply as distinct species of one genus, *Troglodytes;* by others as distinct genera—*Troglodytes* being reserved for the Chimpanzees, and *Gorilla* for the Engé-ena or Pongo.

Sound knowledge respecting the habits and mode of life of the man-like Apes has been even more difficult of attainment than correct information regarding their structure.

Once in a generation, a Wallace may be found physically, mentally, and morally qualified to wander unscathed through the tropical wilds of

America and of Asia; to form magnificent collections as he wanders; and withal to think out sagaciously the conclusions suggested by his collections: but, to the ordinary explorer or collector, the dense forests of equatorial Asia and Africa, which constitute the favourite habitation of the Orang, the Chimpanzee, and the Gorilla, present difficulties of no ordinary magnitude; and the man who risks his life by even a short visit to the malarious shores of those regions may well be excused if he shrinks from facing the dangers of the interior; if he contents himself with stimulating the industry of the better seasoned natives, and collecting and collating the more or less mythical reports and traditions with which they are too ready to supply him.

In such a manner most of the earlier accounts of the habits of the man-like Apes originated; and even now a good deal of what passes current must be admitted to have no very safe foundation. The best information we possess is that, based almost wholly on direct European testimony, respecting the Gibbons; the next best evidence relates to the Orangs; while our knowledge of the habits of the Chimpanzee and the Gorilla stands much in need of support and enlargement by additional testimony from instructed European eye-witnesses.

It will therefore be convenient in endeavouring to form a notion of what we are justified in believing about these animals, to commence with the best known man-like Apes, the Gibbons and Orangs; and to make use of the perfectly trustworthy information respecting them as a sort of criterion of the probable truth or falsehood of assertions respecting the others.

Of the GIBBONS, half a dozen species are found scattered over the Asiatic islands, Java, Sumatra, Borneo, and through Malacca, Siam, Arracan, and an uncertain extent of Hindostan, on the main land of Asia. The largest attain a few inches above three feet in height, from the crown to the heel, so that they are shorter than the other man-like Apes; while the slenderness of their bodies renders their mass far smaller in proportion even to this diminished height.

Dr. Salomon Müller, an accomplished Dutch naturalist, who lived for many years in the Eastern Archipelago, and to the results of whose personal experience I shall frequently have occasion to refer, states that the Gibbons are true mountaineers, loving the slopes and edges of the hills, though they rarely ascend beyond the limit of the fig-trees. All day long they haunt the tops of the tall trees; and though, towards evening, they descend in small troops to the open ground, no sooner do they spy a man than they dart up the hill-sides, and disappear in the darker valleys.

All observers testify to the prodigious volume of

Fig. 8.—A Gibbon (*H. pileatus*), after Wolf.

voice possessed by these animals. According to the writer whom I have just cited, in one of them, the Siamang, "the voice is grave and penetrating, resembling the sounds gōek, gōek, gōek, gōek, goek ha ha ha ha haaāāā, and may easily be heard at a distance of half a league." While the cry is being uttered, the great membranous bag under the throat which communicates with the organ of voice, the so-called "laryngeal sac," becomes greatly distended, diminishing again when the creature relapses into silence.

M. Duvaucel, likewise, affirms that the cry of the Siamang may be heard for miles—making the woods ring again. So Mr. Martin[1] describes the cry of the agile Gibbon as "overpowering and deafening" in a room, and "from its strength, well calculated for resounding through the vast forests." Mr. Waterhouse, an accomplished musician as well as zoologist, says, "The Gibbon's voice is certainly much more powerful than that of any singer I ever heard." And yet it is to be recollected that this animal is not half the height of, and far less bulky in proportion than, a man.

There is good testimony that various species of Gibbon readily take to the erect posture. Mr. George Bennett,[2] a very excellent observer, in describing the habits of a male *Hylobates syndactylus* which remained for some time in his possession,

---

[1] *Man and Monkies*, p. 423.
[2] *Wanderings in New South Wales*, vol. ii. chap. viii. 1834.

says: "He invariably walks in the erect posture when on a level surface; and then the arms either hang down, enabling him to assist himself with his knuckles; or what is more usual, he keeps his arms uplifted in nearly an erect position, with the hands pendent ready to seize a rope, and climb up on the approach of danger or on the obtrusion of strangers. He walks rather quick in the erect posture, but with a waddling gait, and is soon run down if, whilst pursued, he has no opportunity of escaping by climbing . . . . When he walks in the erect posture he turns the leg and foot outwards, which occasions him to have a waddling gait and to seem bow-legged."

Dr. Burrough states of another Gibbon, the Horlack or Hooluk:

"They walk erect; and when placed on the floor, or in an open field, balance themselves very prettily, by raising their hands over their head and slightly bending the arm at the wrist and elbow, and then run tolerably fast, rocking from side to side; and, if urged to greater speed, they let fall their hands to the ground, and assist themselves forward, rather jumping than running, still keeping the body, however, nearly erect."

Somewhat different evidence, however, is given by Dr. Winslow Lewis:[1]

"Their only manner of walking was on their posterior or inferior extremities, the others being raised upwards to preserve their equilibrium, as

[1] *Boston Journal of Natural History*, vol. i. 1834.

rope-dancers are assisted by long poles at fairs. Their progression was not by placing one foot before the other, but by simultaneously using both, as in jumping." Dr. Salomon Müller also states that the Gibbons progress along the ground by short series of tottering jumps, effected only by the hind limbs, the body being held altogether upright.

But Mr. Martin (*l. c.* p. 418), who also speaks from direct observation, says of the Gibbons generally:

"Pre-eminently qualified for arboreal habits, and displaying among the branches amazing activity, the Gibbons are not so awkward or embarrassed on a level surface as might be imagined. They walk erect, with a waddling or unsteady gait, but at a quick pace; the equilibrium of the body requiring to be kept up, either by touching the ground with the knuckles, first on one side then on the other, or by uplifting the arms so as to poise it. As with the Chimpanzee, the whole of the narrow, long sole of the foot is placed upon the ground at once and raised at once, without any elasticity of step."

After this mass of concurrent and independent testimony, it cannot reasonably be doubted that the Gibbons commonly and habitually assume the erect attitude.

But level ground is not the place where these animals can display their very remarkable and peculiar locomotive powers, and that prodigious activity which almost tempts one to rank them among flying, rather than among ordinary climbing mammals.

Mr. Martin (*l. c.* p. 430) has given so excellent and graphic an account of the movements of a *Hylobates agilis*, living in the Zoological Gardens, in 1840, that I will quote it in full:

"It is almost impossible to convey in words an idea of the quickness and graceful address of her movements: they may indeed be termed aerial, as she seems merely to touch in her progress the branches among which she exhibits her evolutions. In these feats her hands and arms are the sole organs of locomotion; her body hanging as if suspended by a rope, sustained by one hand (the right for example), she launches herself, by an energetic movement, to a distant branch, which she catches with the left hand; but her hold is less than momentary: the impulse for the next launch is acquired: the branch then aimed at is attained by the right hand again and quitted instantaneously, and so on in alternate succession. In this manner spaces of twelve and eighteen feet are cleared, with the greatest ease and uninterruptedly, for hours together, without the slightest appearance of fatigue being manifested; and it is evident that if more space could be allowed, distances very greatly exceeding eighteen feet would be as easily cleared; so that Duvaucel's assertion that he had seen these animals launch themselves from one branch to another, forty feet asunder, startling as it is, may be well credited. Sometimes, on seizing a branch in her progress, she will throw herself, by the power of one arm only, completely round it, making a revolution with such rapidity as almost to deceive the eye, and continue her progress with undiminished velocity. It is singular to observe how suddenly this Gibbon can stop, when the impetus given by the rapidity and distance of her swinging leaps would seem to require a gradual abatement of her movements. In the very midst of her flight a branch is seized, the body raised, and she is seen, as if by magic, quietly seated on it, grasping it with her feet. As suddenly she again throws herself into action.

"The following facts will convey some notion of her dexterity and quickness. A live bird was let loose in her apartment; she marked its flight, made a long swing to a distant branch, caught the bird with one hand in her passage, and attained the branch with her other hand; her aim, both at the bird and at the branch, being as successful as if one object only had engaged her attention. It may be added that she instantly bit off the head of the bird, picked its feathers, and then threw it down without attempting to eat it.

"On another occasion this animal swung herself from a perch, across a passage at least twelve feet wide, against a window which it was thought would be immediately broken: but not so; to the surprise of all, she caught the narrow framework between the panes with her hand, in an instant attained the proper impetus, and sprang back again to the cage she had left—a feat requiring not only great strength, but the nicest precision."

The Gibbons appear to be naturally very gentle, but there is very good evidence that they will bite severely when irritated—a female *Hylobates agilis* having so severely lacerated one man with her long canines, that he died; while she had injured others so much that, by way of precaution, these formidable teeth had been filed down; but, if threatened, she would still turn on her keeper The Gibbons eat insects, but appear generally to avoid animal food. A Siamang, however, was seen by Mr. Bennett to seize and devour greedily a live lizard. They commonly drink by dipping their fingers in the liquid and then licking them. It is asserted that they sleep in a sitting posture.

Duvaucel affirms that he has seen the females carry their young to the waterside and there wash their faces, in spite of resistance and cries. They are gentle and affectionate in captivity—full of tricks and pettishness, like spoiled children, and yet not devoid of a certain conscience, as an anecdote, told by Mr. Bennett (*l. c.* p. 156), will show. It would appear that his Gibbon had a peculiar inclination for disarranging things in the cabin. Among these articles, a piece of soap would especially attract his notice, and for the removal of this he had been once or twice scolded. "One morning," says Mr. Bennett, "I was writing, the ape being present in the cabin, when casting my eyes towards him, I saw the little fellow taking the soap. I watched him without his perceiving that I did so: and he occasionally would cast a furtive glance towards the place where I sat. I pretended to write; he, seeing me busily occupied, took the soap, and moved away with it in his paw. When he had walked half the length of the cabin, I spoke quietly, without frightening him. The instant he found I saw him, he walked back again, and deposited the soap nearly in the same place from whence he had taken it. There was certainly something more than instinct in that action: he evidently betrayed a consciousness of having done wrong both by his first and last actions—and what is reason if that is not an exercise of it?"

The most elaborate account of the natural

Fig. 9.—An adult male Orang-Utan, after Müller and Schlegel.

history of the ORANG-UTAN extant, is that given in the "Verhandelingen over de Natuurlijke Geschiedenis der Nederlandsche overzeesche Bezittingen (1839-45)," by Dr. Salomon Müller and Dr. Schlegel, and I shall base what I have to say upon this subject almost entirely on their statements, adding, here and there, particulars of interest from the writings of Brooke, Wallace, and others.

The Orang-Utan would rarely seem to exceed four feet in height, but the body is very bulky, measuring two-thirds of the height in circumference.[1]

The Orang-Utan is found only in Sumatra and Borneo, and is common in neither of these islands—in both of which it occurs always in low, flat plains, never in the mountains. It loves the densest and most sombre of the forests, which

---

[1] The largest Orang-Utan, cited by Temminck, measured, when standing upright, four feet; but he mentions having just received news of the capture of an Orang five feet three inches high. Schlegel and Müller say that their largest old male measured, upright, 1.25 Netherlands "el"; and from the crown to the end of the toes, 1.5 el; the circumference of the body being about 1 el. The largest old female was 1.09 el high, when standing. The adult skeleton in the College of Surgeons' Museum, if set upright, would stand 3 ft. 6-8 in. from crown to sole. Dr. Humphry gives 3 ft. 8 in. as the mean height of two Orangs. Of seventeen Orangs examined by Mr. Wallace, the largest was 4 ft. 2 in. high, from the heel to the crown of the head. Mr. Spencer St. John, however, in his *Life in the Forests of the Far East*, tells us of an Orang of "5 ft. 2 in., measuring fairly from the head to the heel," 15 in. across the face, and 12 in. round the wrist. It does not appear, however, Mr. that St. John measured this Orang himself.

extend from the sea-shore inland, and thus is found only in the eastern half of Sumatra, where alone such forests occur, though, occasionally, it strays over to the western side.

On the other hand, it is generally distributed through Borneo, except in the mountains, or where the population is dense. In favourable places, the hunter may, by good fortune, see three or four in a day.

Except in the pairing time, the old males usually live by themselves. The old females, and the immature males, on the other hand, are often met with in twos and threes; and the former occasionally have young with them, though the pregnant females usually separate themselves, and sometimes remain apart after they have given birth to their offspring. The young Orangs seem to remain unusually long under their mother's protection, probably in consequence of their slow growth. While climbing, the mother always carries her young against her bosom, the young holding on by his mother's hair.[1] At what time of life the Orang-Utan becomes capable of propagation, and how long the females go with young, is unknown, but it is probable that they are not adult until they

---

[1] See Mr. Wallace's account of an infant "Orang-utan," in the *Annals of Natural History* for 1856. Mr. Wallace provided his interesting charge with an artificial mother of buffalo-skin, but the cheat was too successful. The infant's entire experience led it to associate teats with hair, and feeling the latter, it spent its existence in vain endeavours to discover the former.

arrive at ten or fifteen years of age. A female which lived for five years at Batavia, had not attained one-third the height of the wild females. It is probable that, after reaching adult years, they go on growing, though slowly, and that they live to forty or fifty years. The Dyaks tell of old Orangs, which have not only lost all their teeth, but which find it so troublesome to climb, that they maintain themselves on windfalls and juicy herbage.

The Orang is sluggish, exhibiting none of that marvellous activity characteristic of the Gibbons. Hunger alone seems to stir him to exertion, and when it is stilled, he relapses into repose. When the animal sits, it curves its back and bows its head, so as to look straight down on the ground; sometimes it holds on with its hands by a higher branch, sometimes lets them hang phlegmatically down by its side—and in these positions the Orang will remain, for hours together, in the same spot, almost without stirring, and only now and then giving utterance to his deep, growling voice. By day, he usually climbs from one tree-top to another, and only at night descends to the ground, and if then threatened with danger, he seeks refuge among the underwood. When not hunted, he remains a long time in the same locality, and sometimes stops for many days on the same tree —a firm place among its branches serving him for a bed. It is rare for the Orang to pass the night in the summit of a large tree, probably because it is too windy and cold there for him; but, as soon as night draws on, he descends from the height and seeks out a fit bed in the lower and darker part, or in the leafy top of a small tree, among which he prefers Nibong Palms, Pandani, or one of those parasitic Orchids which give the primæval forests of Borneo so characteristic and striking an appearance. But wherever he determines to sleep, there he prepares himself a sort of nest: little boughs and leaves are drawn together round the selected spot, and bent crosswise over one another; while to make the bed soft, great leaves of Ferns, of Orchids, of *Pandanus fascicularis*, *Nipa fruticans*, &c., are laid over them. Those which Müller saw, many of them being very fresh, were situated at a height of ten to twenty-five feet above the ground, and had a circumference, on the average, of two or three feet. Some were packed many inches thick with *Pandanus* leaves; others were remarkable only for the cracked twigs, which, united in a common centre, formed a regular platform. "The rude *hut*," says Sir James Brooke, "which they are stated to build in the trees, would be more properly called a seat or nest, for it has no roof or cover of any sort. The facility with which they form this nest is curious, and I had an opportunity of seeing a wounded female weave the branches together and seat herself, within a minute."

According to the Dyaks the Orang rarely leaves

his bed before the sun is well above the horizon and has dissipated the mists. He gets up about nine, and goes to bed again about five; but sometimes not till late in the twilight. He lies sometimes on his back; or, by way of change, turns on one side or the other, drawing his limbs up to his body, and resting his head on his hand. When the night is cold, windy, or rainy, he usually covers his body with a heap of *Pandanus*, *Nipa*, or Fern leaves, like those of which his bed is made, and he is especially careful to wrap up his head in them. It is this habit of covering himself up which has probably led to the fable that the Orang builds huts in the trees.

Although the Orang resides mostly amid the boughs of great trees, during the daytime, he is very rarely seen squatting on a thick branch, as other apes, and particularly the Gibbons, do. The Orang, on the contrary, confines himself to the slender leafy branches, so that he is seen right at the top of the trees, a mode of life which is closely related to the constitution of his hinder limbs, and especially to that of his seat. For this is provided with no callosities, such as are possessed by many of the lower apes, and even by the Gibbons; and those bones of the pelvis, which are termed the ischia, and which form the solid framework of the surface on which the body rests in the sitting posture, are not expanded like those of the apes which possess callosities, but are more like those of man.

An Orang climbs so slowly and cautiously,[1] as, in this act, to resemble a man more than an ape, taking great care of his feet, so that injury of them seems to affect him far more than it does other apes. Unlike the Gibbons, whose forearms do the greater part of the work, as they swing from branch to branch, the Orang never makes even the smallest jump. In climbing, he moves alternately one hand and one foot, or, after having laid fast hold with the hands, he draws up both feet together. In passing from one tree to another, he always seeks out a place where the twigs of both come close together, or interlace. Even when closely pursued, his circumspection is amazing: he shakes the branches to see if they will bear him, and then bending an overhanging bough down by throwing his weight gradually along it, he makes a bridge from the tree he wishes to quit to the next.[1]

On the ground the Orang always goes laboriously and shakily, on all fours. At starting he will run faster than a man, though he may soon be overtaken. The very long arms which, when

---

[1] "They are the slowest and least active of all the monkey tribe, and their motions are surprisingly awkward and uncouth."—Sir James Brooke, in the *Proceedings of the Zoological Society*, 1841.

[2] Mr. Wallace's account of the progression of the Orang almost exactly corresponds with this.

he runs, are but little bent, raise the body of the Orang remarkably, so that he assumes much the posture of a very old man bent down by age, and making his way along by the help of a stick. In walking, the body is usually directed straight forward, unlike the other apes, which run more or less obliquely; except the Gibbons, who in these as in so many other respects, depart remarkably from their fellows.

The Orang cannot put its feet flat on the ground, but is supported upon their outer edges, the heel resting more on the ground, while the curved toes partly rest upon the ground by the upper side of their first joint, the two outermost toes of each foot completely resting on this surface. The hands are held in the opposite manner, their inner edges serving as the chief support. The fingers are then bent out in such a manner that their foremost joints, especially those of the two innermost fingers, rest upon the ground by their upper sides, while the point of the free and straight thumb serves as an additional fulcrum.

The Orang never stands on its hind legs, and all the pictures, representing it as so doing, are as false as the assertion that it defends itself with sticks, and the like.

The long arms are of especial use, not only in climbing, but in the gathering of food from boughs to which the animal could not trust his weight. Figs, blossoms, and young leaves of various kinds, constitute the chief nutriment of the Orang; but strips of bamboo two or three feet long were found in the stomach of a male. They are not known to eat living animals.

Although, when taken young, the Orang-Utan soon becomes domesticated, and indeed seems to court human society, it is naturally a very wild and shy animal, though apparently sluggish and melancholy. The Dyaks affirm, that when the old males are wounded with arrows only, they will occasionally leave the trees and rush raging upon their enemies, whose sole safety lies in instant flight, as they are sure to be killed if caught.[1]

[1] Sir James Brooke, in a letter to Mr. Waterhouse, published in the proceedings of the Zoological Society for 1841, says:— "On the habits of the Orangs, as far as I have been able to observe them, I may remark that they are as dull and slothful as can well be conceived, and on no occasion, when pursuing them, did they move so fast as to preclude my keeping pace with them easily through a moderately clear forest; and even when obstructions below (such as wading up to the neck) allowed them to get away some distance, they were sure to stop and allow me to come up. I never observed the slightest attempt at defence, and the wood which sometimes rattled about our ears was broken by their weight, and not thrown, as some persons represent. If pushed to extremity, however, the *Pappan* could not be otherwise than formidable, and one unfortunate man, who, with a party, was trying to catch a large one alive, lost two of his fingers, besides being severely bitten on the face, whilst the animal finally beat off his pursuers and escaped."

Mr. Wallace, on the other hand, affirms that he has several times observed them throwing down branches when pursued. "It is true he does not throw them *at* a person, but casts them down vertically; for it is evident that a bough cannot be thrown to any distance from the top of a lofty tree. In one case a female Mias, on a durian tree, kept up for at least ten minutes a continuous shower of branches and of the heavy, spined fruits,

But, though possessed of immense strength, it is rare for the Orang to attempt to defend itself, especially when attacked with fire-arms. On such occasions he endeavours to hide himself, or to escape along the topmost branches of the trees, breaking off and throwing down the boughs as he goes. When wounded he betakes himself to the highest attainable point of the tree, and emits a singular cry, consisting at first of high notes, which at length deepen into a low roar, not unlike that of a panther. While giving out the high notes the Orang thrusts out his lips into a funnel shape; but in uttering the low notes he holds his mouth wide open, and at the same time the great throat bag, or laryngeal sac, becomes distended.

According to the Dyaks, the only animal the Orang measures his strength with is the crocodile, who occasionally seizes him on his visits to the water side. But they say that the Orang is more than a match for his enemy, and beats him to death, or rips up his throat by pulling the jaws asunder!

Much of what has been here stated was

as large as 32-pounders, which most effectually kept us clear of the tree she was on. She could be seen breaking them off and throwing them down with every appearance of rage, uttering at intervals a loud pumping grunt, and evidently meaning mischief."—" On the Habits of the Orang-Utan," *Annals of Natural History.* 1856. This statement, it will be observed, is quite in accordance with that contained in the letter of the Resident Palm quoted above (p. 23).

probably derived by Dr. Müller from the reports of his Dyak hunters; but a large male, four feet high, lived in captivity, under his observation, for a month, and receives a very bad character.

"He was a very wild beast," says Müller, "of prodigious strength, and false and wicked to the last degree. If any one approached he rose up slowly with a low growl, fixed his eyes in the direction in which he meant to make his attack, slowly passed his hand between the bars of his cage, and then extending his long arm, gave a sudden grip—usually at the face." He never tried to bite (though Orangs will bite one another), his great weapons of offence and defence being his hands.

His intelligence was very great; and Müller remarks that though the faculties of the Orang have been estimated too highly, yet Cuvier, had he seen this specimen, would not have considered its intelligence to be only a little higher than that of the dog.

His hearing was very acute, but the sense of vision seemed to be less perfect. The under lip was the great organ of touch, and played a very important part in drinking, being thrust out like a trough, so as either to catch the falling rain, or to receive the contents of the half cocoa-nut shell full of water with which the Orang was supplied, and which, in drinking, he poured into the trough thus formed.

In Borneo the Orang-Utan of the Malays goes by the name of "*Mias*" among the Dyaks, who distinguish several kinds as *Mias Pappan*, or *Zimo, Mias Kassu,* and *Mias Rambi.* Whether these are distinct species, however, or whether they are mere races, and how far any of them are identical with the Sumatran Orang, as Mr. Wallace thinks the Mias Pappan to be, are problems which are at present undecided; and the variability of these great apes is so extensive, that the settlement of the question is a matter of great difficulty. Of the form called "Mias Pappan," Mr. Wallace [1] observes,

"It is known by its large size, and by the lateral expansion of the face into fatty protuberances, or ridges, over the temporal muscles, which have been mis-termed *callosities*, as they are perfectly soft, smooth, and flexible. Five of this form, measured by me, varied only from 4 feet 1 inch to 4 feet 2 inches in height, from the heel to the crown of the head, the girth of the body from 3 feet to 3 feet 7½ inches, and the extent of the outstretched arms from 7 feet 2 inches to 7 feet 6 inches; the width of the face from 10 to 13¼ inches. The colour and length of the hair varied in different individuals, and in different parts of the same individual; some possessed a rudimentary nail on the great toe, others none at all; but they otherwise present no external differences on which to establish even varieties of a species.

"Yet, when we examine the crania of these individuals, we find remarkable differences of form, proportion, and dimension, no two being exactly alike. The slope of the profile, and the projection of the muzzle, together with the size of the cranium,

---

[1] On the Orang-Utan, or Mias of Borneo, *Annals of Natural History*, 1856.

offer differences as decided as those existing between the most strongly marked forms of the Caucasian and African crania in the human species. The orbits vary in width and height, the cranial ridge is either single or double, either much or little developed, and the zygomatic aperture varies considerably in size. This variation in the proportions of the crania enables us satisfactorily to explain the marked difference presented by the single-crested and double-crested skulls, which have been thought to prove the existence of two large species of Orang. The external surface of the skull varies considerably in size, as do also the zygomatic aperture and the temporal muscle; but they bear no necessary relation to each other, a small muscle often existing with a large cranial surface, and *vice versâ*. Now, those skulls which have the largest and strongest jaws and the widest zygomatic aperture, have the muscles so large that they meet on the crown of the skull, and deposit the bony ridge which separates them, and which is the highest in that which has the smallest cranial surface. In those which combine a large surface with comparatively weak jaws, and small zygomatic aperture, the muscles, on each side, do not extend to the crown, a space of from 1 to 2 inches remaining between them, and along their margins small ridges are formed. Intermediate forms are found, in which the ridges meet only in the hinder part of the skull. The form and size of the ridges are therefore independent of age, being sometimes more strongly developed in the less aged animal. Professor Temminck states that the series of skulls in the Leyden Museum shows the same result."

Mr. Wallace observed two male adult Orangs (Mias Kassu of the Dyaks), however, so very different from any of these that he concludes them to be specifically distinct; they were respectively 3 feet 8½ inches and 3 feet 9½ inches high, and possessed no sign of the cheek excrescences, but otherwise resembled the larger kinds. The skull has no crest, but two bony ridges, 1¾ inches to 2 inches apart, as in the *Simia morio* of Professor Owen. The teeth, however, are immense, equalling or surpassing those of the other species. The females of both these kinds, according to Mr. Wallace, are devoid of excrescences, and resemble the smaller males, but are shorter by 1½ to 3 inches, and their canine teeth are comparatively small, subtruncated and dilated at the base, as in the so-called *Simia morio*, which is, in all probability, the skull of a female of the same species as the smaller males. Both males and females of this smaller species are distinguishable, according to Mr. Wallace, by the comparatively large size of the middle incisors of the upper jaw.

So far as I am aware, no one has attempted to dispute the accuracy of the statements which I have just quoted regarding the habits of the two Asiatic man-like apes; and if true, they must be admitted as evidence, that such an Ape—

1stly, May readily move along the ground in the erect, or semi-erect, position, and without direct support from its arms.

2ndly, That it may possess an extremely loud voice, so loud as to be readily heard one or two miles.

3rdly, That it may be capable of great viciousness and violence when irritated: and this is especially true of adult males.

4thly, That it may build a nest to sleep in.

Such being well established facts respecting the Asiatic Anthropoids, analogy alone might justify us in expecting the African species to offer similar peculiarities, separately or combined; or, at any rate, would destroy the force of any attempted *a priori* argument against such direct testimony as might be adduced in favour of their existence. And, if the organization of any of the African Apes could be demonstrated to fit it better than either of its Asiatic allies for the erect position and for efficient attack, there would be still less reason for doubting its occasional adoption of the upright attitude or of aggressive proceedings.

From the time of Tyson and Tulpius downwards, the habits of the young CHIMPANZEE in a state of captivity have been abundantly reported and commented upon. But trustworthy evidence as to the manners and customs of adult anthropoids of this species, in their native woods, was almost wanting up to the time of the publication of the paper by Dr. Savage, to which I have already referred; containing notes of the observations which he made, and of the information which he collected from sources which he considered trustworthy, while resident at Cape Palmas, at the north-western limit of the Bight of Benin.

The adult Chimpanzees measured by Dr. Savage, never exceeded, though the males may almost attain, five feet in height.

"When at rest the sitting posture is that generally assumed. They are sometimes seen standing and walking, but when thus detected, they immediately take to all fours, and flee from the presence of the observer. Such is their organisation that they cannot stand erect, but lean forward. Hence they are seen, when standing, with the hands clasped over the occiput, or the lumbar region, which would seem necessary to balance or ease of posture.

"The toes of the adult are strongly flexed and turned inwards, and cannot be perfectly straightened. In the attempt the skin gathers into thick folds on the back, showing that the full expansion of the foot, as is necessary in walking, is unnatural. The natural position is on all fours, the body anteriorly resting upon the knuckles. These are greatly enlarged, with the skin protuberant and thickened like the sole of the foot.

"They are expert climbers, as one would suppose from their organisation. In their gambols they swing from limb to limb to a great distance, and leap with astonishing agility. It is not unusual to see the 'old folks' (in the language of an observer) sitting under a tree regaling themselves with fruit and friendly chat, while their 'children' are leaping around them, and swinging from tree to tree with boisterous merriment.

"As seen here, they cannot be called *gregarious*, seldom more than five, or ten at most, being found together. It has been said, on good authority, that they occasionally assemble in large numbers, in gambols. My informant asserts that he saw once not less than fifty so engaged; hooting, screaming, and drumming with sticks upon old logs, which is done in the latter case with equal facility by the four extremities. They do not appear ever to act on the offensive, and seldom, if ever really, on the defensive. When about to be captured, they resist by throwing their arms about their opponent, and attempting to draw him into contact with their teeth." (Savage, *l.c.* p. 384.)

With respect to this last point Dr. Savage is very explicit in another place :

"*Biting* is their principal art of defence. I have seen one man who had been thus severely wounded in the feet.

"The strong development of the canine teeth in the adult would seem to indicate a carnivorous propensity; but in no state save that of domestication do they manifest it. At first they reject flesh, but easily acquire a fondness for it. The canines are early developed, and evidently designed to act the important part of weapons of defence. When in contact with man almost the first effort of the animal is—*to bite.*

"They avoid the abodes of men, and build their habitations in trees. Their construction is more that of *nests* than *huts*, as they have been erroneously termed by some naturalists. They generally build not far above the ground. Branches or twigs are bent, or partly broken, and crossed, and the whole supported by the body of a limb or a crotch. Sometimes a nest will be found near the *end* of a *strong leafy branch* twenty or thirty feet from the ground. One I have lately seen that could not be less than forty feet, and more probably it was fifty. But this is an unusual height.

"Their dwelling-place is not permanent, but changed in pursuit of food and solitude, according to the force of circumstances. We more often see them in elevated places; but this arises from the fact that the low grounds, being more favourable for the natives' rice-farms, are the oftener cleared, and hence are almost always wanting in suitable trees for their nests. . . . It is seldom that more than one or two nests are seen upon the same tree, or in the same neighbourhood: five have been found, but it was an unusual circumstance." . . .

"They are very filthy in their habits. . . . It is a tradition with the natives generally here, that they were once members of their own tribe: that for their depraved habits they were expelled from all human society, and, that through an obstinate indulgence of their vile propensities, they have degenerated into their present state and organisation. They are, however, eaten by them, and when cooked with the oil and pulp of the palm-nut considered a highly palatable morsel.

"They exhibit a remarkable degree of intelligence in their habits, and, on the part of the mother, much affection for their young. The second female described was upon a tree when first discovered, with her mate and two young ones (a male and a female). Her first impulse was to descend with great rapidity and make off into the thicket, with her mate and female offspring. The young male remaining behind, she soon returned to the rescue. She ascended and took him in her arms, at which moment she was shot, the ball passing through the fore-arm of the young one, on its way to the heart of the mother. . . .

"In a recent case, the mother, when discovered, remained upon the tree with her offspring, watching intently the movements of the hunter. As he took aim, she motioned with her hand, precisely in the manner of a human being, to have him desist and go away. When the wound has not proved instantly fatal, they have been known to stop the flow of blood by pressing with the hand upon the part, and when this did not succeed, to apply leaves and grass . . . . When shot, they give a sudden screech, not unlike that of a human being in sudden and acute distress."

The ordinary voice of the Chimpanzee, however, is affirmed to be hoarse, guttural, and not very loud, somewhat like "whoo-whoo." (*l. c.* p. 365.)

The analogy of the Chimpanzee to the Orang, in its nest-building habit and in the mode of forming its nest, is exceedingly interesting; while, on the other hand, the activity of this ape, and its tendency to bite, are particulars in which it rather resembles the Gibbons. In extent of geographical range, again, the Chimpanzees—which are found from Sierra Leone to Congo—remind one of the Gibbons, rather than of either of the other man-like apes; and it seems not unlikely that, as is the case with the Gibbons, there may be several species spread over the geographical area of the genus.

The same excellent observer, from whom I have borrowed the preceding account of the habits of the adult Chimpanzee, published fifteen years ago,[1] an account of the GORILLA, which has, in its most essential points, been confirmed by subsequent observers, and to which so very little has really been added, that in justice to Dr. Savage I give it almost in full.

"It should be borne in mind that my account is based upon the statements of the aborigines of that region (the Gaboon). In this connection, it may also be proper for me to remark, that having been a missionary resident for several years, studying, from habitual intercourse, the African mind and character, I felt myself prepared to discriminate and decide upon the probability of their statements. Besides, being familiar with the history and habits of its interesting congener (*Trog. niger*, Geoff.), I was able to separate their accounts of the two animals, which, having the same locality and a similarity of habit, are confounded in the minds of the mass, especially as but few—such as traders to the interior and huntsmen—have ever seen the animal in question.

"The tribe from which our knowledge of the animal is derived, and whose territory forms its habitat, is the *Mpongwe*, occupying both banks of the River Gaboon, from its mouth to some fifty or sixty miles upward. . . .

"If the word 'Pongo' be of African origin, it is probably a corruption of the word *Mpongwe*, the name of the tribe on the banks of the Gaboon, and hence applied to the region they inhabit. Their local name for the Chimpanzee is *Enché-eko*, as

---

[1] Notice of the external characters and habits of Troglodytes Gorilla. *Boston Journal of Natural History*, 1847.

Fig. 10.—The Gorilla, after Wolf.

near as it can be Anglicised, from which the common term 'Jocko' probably comes. The Mpongwe appellation for its new congener is *Engé-ena*, prolonging the sound of the first vowel, and slightly sounding the second.

"The habitat of the *Engé-ena* is the interior of lower Guinea, whilst that of the *Enché-eko* is nearer the sea-board.

"Its height is about five feet; it is disproportionately broad across the shoulders, thickly covered with coarse black hair, which is said to be similar in its arrangement to that of the *Enché-eko;* with age it becomes gray, which fact has given rise to the report that both animals are seen of different colours.

"*Head.*—The prominent features of the head are, the great width and elongation of the face, the depth of the molar region, the branches of the lower jaw being very deep and extending far backward, and the comparative smallness of the cranial portion; the eyes are very large, and said to be like those of the Enché-eko, a bright hazel; nose broad and flat, slightly elevated towards the root; the muzzle broad, and prominent lips and chin, with scattered gray hairs; the under lip highly mobile, and capable of great elongation when the animal is enraged, then hanging over the chin; skin of the face and ears naked, and of a dark brown, approaching to black.

"The most remarkable feature of the head is a high ridge, or crest of hair, in the course of the sagittal suture, which meets posteriorily with a transverse ridge of the same, but less prominent, running round from the back of one ear to the other. The animal has the power of moving the scalp freely forward and back, and when enraged is said to contract it strongly over the brow, thus bringing down the hairy ridge and pointing the hair forward, so as to present an indescribably ferocious aspect.

"Neck short, thick, and hairy; chest and shoulders very broad, said to be fully double the size of the Enché-ekos; arms very long, reaching some way below the knee—the fore-arm much the shortest; hands very large, the thumbs much larger than the fingers. . . .

"The gait is shuffling; the motion of the body, which is never upright as in man, but bent forward, is somewhat rolling, or from side to side. The arms being longer than the Chimpanzee, it does not stoop as much in walking; like that animal, it makes progression by thrusting its arms forward, resting the hands on the ground, and then giving the body a half jumping, half swinging motion between them. In this act it is said not to flex the fingers, as does the Chimpanzee, resting on its knuckles, but to extend them, making a fulcrum of the hand. When it assumes the walking posture, to which it is said to be much inclined, it balances its huge body by flexing its arms upward.

Fig. 11.—Gorilla walking (after Wolff).

"They live in bands, but are not so numerous as the Chimpanzees; the females generally exceed the other sex in number. My informants all agree in the assertion that but one adult male is seen in a band; that when the young males grow up, a contest takes place for mastery, and the strongest, by killing and driving out the others, establishes himself as the head of the community."

Dr. Savage repudiates the stories about the Gorillas carrying off women and vanquishing elephants and then adds—

"Their dwellings, if they may be so called, are similar to those of the Chimpanzee, consisting simply of a few sticks and leafy branches, supported by the crotches and limbs of trees: they afford no shelter, and are occupied only at night.

"They are exceedingly ferocious, and always offensive in their habits, never running from man, as does the Chimpanzee. They are objects of terror to the natives, and are never encountered by them except on the defensive. The few that have been captured were killed by elephant hunters and native traders, as they came suddenly upon them while passing through the forests.

"It is said that when the male is first seen he gives a terrific yell, that resounds far and wide through the forest, something like kh—ah! kh—ah! prolonged and shrill. His enormous jaws are widely opened at each expiration, his under lip hangs over the chin, and the hairy ridge and scalp are contracted upon the brow, presenting an aspect of indescribable ferocity.

"The females and young, at the first cry, quickly disappear. He then approaches the enemy in great fury, pouring out his horrid cries in quick succession. The hunter awaits his approach with his gun extended; if his aim is not sure, he permits the animal to grasp the barrel, and as he carries it to his mouth (which is his habit) he fires. Should the gun fail to go off, the barrel (that of the ordinary musket, which is thin) is crushed between his teeth, and the encounter soon proves fatal to the hunter.

"In the wild state, their habits are in general like those of the *Troglodytes niger*, building their nests loosely in trees, living on similar fruits, and changing their place of resort from force of circumstances."

Dr. Savage's observations were confirmed and supplemented by those of Mr. Ford, who communicated an interesting paper on the Gorilla to the Philadelphian Academy of Sciences, in 1852. With respect to the geographical distribution of this greatest of all the man-like Apes, Mr. Ford remarks:

"This animal inhabits the range of mountains that traverse the interior of Guinea, from the Cameroon in the north, to Angola in the south, and about 100 miles inland, and called by the geographers Crystal Mountains. The limit to which this animal extends, either north or south, I am unable to define. But that limit is doubtless some distance north of this river [Gaboon]. I was able to certify myself of this fact in a late excursion to the head-waters of the Mooney (Danger) River, which comes into the sea some sixty miles from this place. I was informed (credibly, I think) that they were numerous among the mountains in which that river rises, and far north of that.

"In the south, this species extends to the Congo River, as I am told by native traders who have visited the coast between the Gaboon and that river. Beyond that, I am not informed. This animal is only found at a distance from the coast in most cases, and, according to my best information, approaches it nowhere so nearly as on the south side of this river, where they have been found within ten miles of the sea. This, however, is only of late occurrence. I am informed by some of the oldest Mpongwe men that formerly he was only found on the sources of the river, but that at present he may be found within half-a-day's walk of its mouth. Formerly he inhabited the mountainous ridge where Bushmen alone inhabited, but now he boldly approaches the Mpongwe plantations. This is doubtless the reason of the scarcity of information in years past, as the opportunities for receiving a knowledge of the animal have not been wanting; traders having for one hundred years frequented this river, and specimens, such as have been brought here within a year, could not have been exhibited without having attracted the attention of the most stupid."

One specimen Mr. Ford examined weighed 170lbs., without the thoracic, or pelvic, viscera, and measured four feet four inches round the chest. This writer describes so minutely and graphically the onslaught of the Gorilla—though he does not for a moment pretend to have witnessed the scene—that I am tempted to give this part of his paper in full, for comparison with other narratives:

"He always rises to his feet when making an attack, though he approaches his antagonist in a stooping posture.

"Though he never lies in wait, yet, when he hears, sees, or scents a man, he immediately utters his characteristic cry, prepares for an attack, and always acts on the offensive. The cry he utters resembles a grunt more than a growl, and is similar to the cry of the Chimpanzee, when irritated, but vastly louder. It is said to be audible at a great distance. His preparation consists in attending the females and young ones, by whom he is usually accompanied, to a little distance. He, however, soon returns, with his crest erect and projecting forward, his nostrils dilated, and his under-lip thrown down, at the same time uttering his characteristic yell, designed, it would seem, to terrify his antagonist. Instantly, unless he is disabled by a well-directed shot, he makes an onset, and, striking his antagonist with the palm of his hands, or seizing him with a grasp from which there is no escape, he dashes him upon the ground, and lacerates him with his tusks.

"He is said to seize a musket, and instantly crush the barrel between his teeth. . . . . . This animal's savage nature is very well shown by the implacable desperation of a young one that was brought here. It was taken very young, and kept four months, and many means were used to tame it; but it was incorrigible, so that it bit me an hour before it died."

Mr. Ford discredits the house-building and elephant-driving stories, and says that no well-informed natives believe them. They are tales told to children.

I might quote other testimony to a similar effect, but, as it appears to me, less carefully weighed and sifted, from the letters of MM. Franquet and Gautier Laboullay, appended to the memoir of M. I. G. St. Hilaire, which I have already cited.

Bearing in mind what is known regarding the Orang and the Gibbon, the statements of Dr. Savage and Mr. Ford do not appear to me to be justly open to criticism on *à priori* grounds. The Gibbons, as we have seen, readily assume the erect posture, but the Gorilla is far better fitted by its organization for that attitude than are the Gibbons: if the laryngeal pouches of the Gibbons, as is very likely, are important in giving volume to a voice which can be heard for half a league, the Gorilla, which has similar sacs, more largely developed, and whose bulk is fivefold that of a Gibbon, may well be audible for twice that distance. If the Orang fights with its hands, the Gibbons and Chimpanzees with their teeth, the Gorilla may, probably enough, do either or both; nor is there anything to be said against either Chimpanzee or Gorilla building a nest, when it is proved that the Orang-Utan habitually performs that feat.

With all this evidence, now ten to fifteen years old, before the world, it is not a little surprising that the assertions of a recent traveller, who, so far as the Gorilla is concerned, really does very little more than repeat, on his own authority, the statements of Savage and of Ford, should have met with so much and such bitter opposition. If subtraction be made of what was known before, the sum and substance of what M. Du Chaillu has affirmed as a matter of his own observation respecting the Gorilla, is, that, in advancing to the attack, the great brute beats his chest with his fists. I confess I see nothing very improbable, or very much worth disputing about, in this statement.

With respect to the other man-like Apes of Africa, M. Du Chaillu tells us absolutely nothing, of his own knowledge, regarding the common Chimpanzee; but he informs us of a bald-headed species or variety, the *nschiego mbouve*, which builds itself a shelter, and of another rare kind with a comparatively small face, large facial angle, and peculiar note, resembling "Kooloo."

As the Orang shelters itself with a rough coverlet of leaves, and the common Chimpanzee, according to that eminently trustworthy observer Dr. Savage, makes a sound like "Whoo-whoo,"— the grounds of the summary repudiation with which M. Du Chaillu's statements on these matters have been met are not obvious.

If I have abstained from quoting M. Du Chaillu's work, then, it is not because I discern any inherent improbability in his assertions respecting the man-like Apes; nor from any wish to throw suspicion on his veracity; but because, in my opinion, so long as his narrative remains in its present state of unexplained and apparently inexplicable confusion, it has no claim to original authority respecting any subject whatsoever.

It may be truth, but it is not evidence.

African Cannibalism in the Sixteenth Century.

In turning over Pigafetta's version of the narrative of Lopez, which I have quoted above, I came upon so curious and unexpected an anticipation, by some two centuries and a half, of one of the most startling parts of M. Du Chaillu's narrative, that I cannot refrain from drawing attention to it in a note, although I must confess that the subject is not strictly relevant to the matter in hand.

In the fifth chapter of the first book of the "Descriptio," "Concerning the northern part of the Kingdom of Congo and its boundaries," is mentioned a people whose king is called "Maniloango," and who live under the equator, and as far westward as Cape Lopez. This appears to be the country now inhabited by the Ogobai and Bakalai according to M. Du Chaillu.—"Beyond these dwell another people called 'Anziques,' of incredible ferocity, for they eat one another, sparing neither friends nor relations."

These people are armed with small bows bound tightly round with snake skins, and strung with a reed or rush. Their arrows, short and slender, but made of hard wood, are shot with great rapidity. They have iron axes, the handles of which are bound round with snake skins, and swords with scabbards of the same material; for defensive armour they employ elephant hides. They cut their skins when young, so as to produce scars. "Their butchers' shops are filled with human flesh instead of that of oxen or sheep. For they eat the enemies whom they take in battle. They fatten, slay and devour their slaves also, unless

74 AFRICAN CANNIBALISM

FIG. 12.—Butcher's Shop of the Anziques Anno 1598.

they think they shall get a good price for them; and, moreover, sometimes for weariness of life or desire of glory (for they think it a great thing and the sign of a generous soul to despise life), or for love of their rulers, offer themselves up for food."

"There are indeed many cannibals, as in the Eastern Indies and in Brazil and elsewhere, but none such as these, since the others only eat their enemies, but these their own blood relations."

The careful illustrators of Pigafetta have done their best to enable the reader to realize this account of the "Anziques," and the unexampled butcher's shop represented in Fig. 12, is a facsimile of part of their Plate XII.

M. Du Chaillu's account of the Fans accords most singularly with what Lopez here narrates of the Anziques. He speaks of their small crossbows and little arrows, of their axes and knives, "ingeniously sheathed in snake skins." "They tattoo themselves more than any other tribes I have met north of the equator." And all the world knows what M. Du Chaillu says of their cannibalism—"Presently we passed a woman who solved all doubt. She bore with her a piece of the thigh of a human body, just as we should go to market and carry thence a roast or steak." M. Du Chaillu's artist cannot generally be accused of any want of courage in embodying the statements of his author, and it is to be regretted that, with so good an excuse, he has not furnished us with a fitting companion to the sketch of the brothers De Bry.

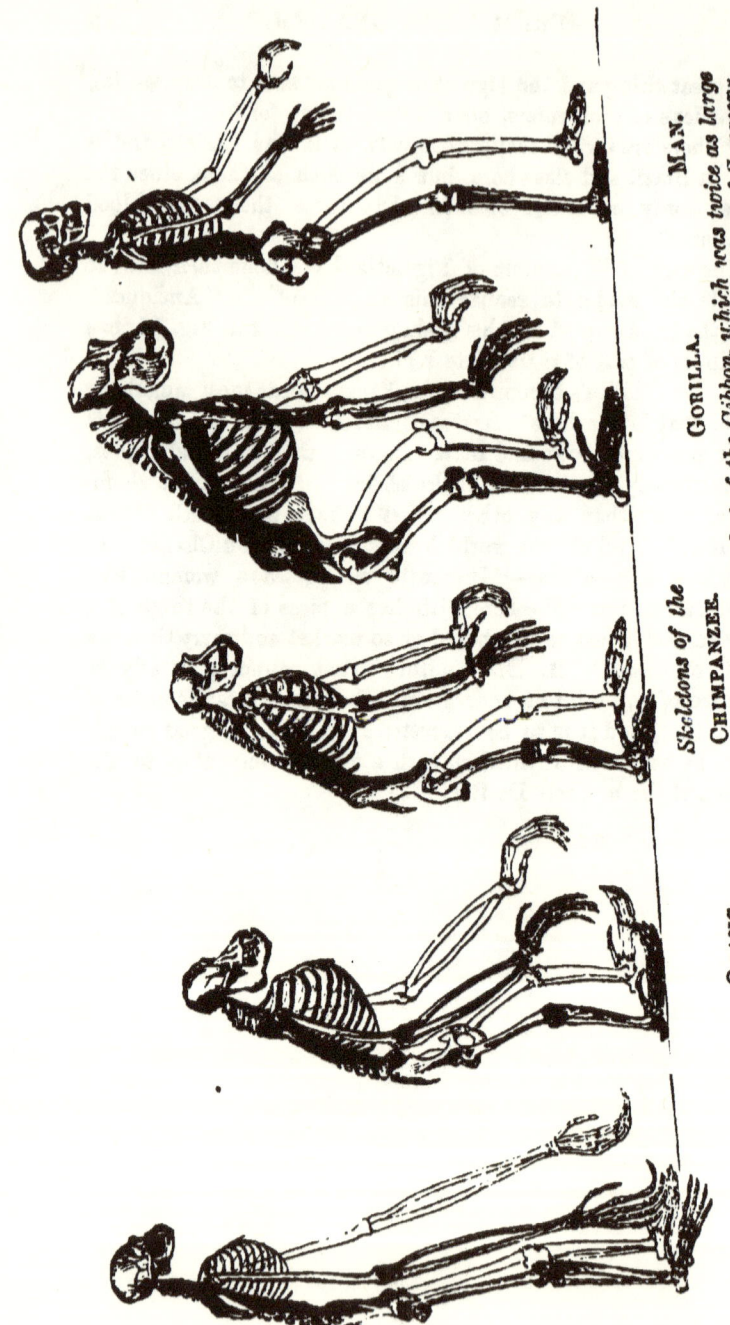

Skeletons of the
GIBBON.   ORANG.   CHIMPANZEE.   GORILLA.   MAN.
Photographically reduced from Diagrams of the natural size (except that of the Gibbon, which was twice as large as nature), drawn by Mr. Waterhouse Hawkins from specimens in the Museum of the Royal College of Surgeons.

II

ON THE RELATIONS OF MAN TO THE LOWER ANIMALS

Multis videri poterit, majorem esse differentiam Simiæ et Hominis, quam diei et noctis; verum tamen hi, comparatione instituta inter summos Europæ Heroës et Hottentottes ad Caput bonæ spei degentes, difficillime sibi persuadebunt, has eosdem habere natales; vel si virginem nobilem aulicam, maxime comtam et humanissimam, conferre vellent cum homine sylvestri et sibi relicto, vix augurari possent, hunc et illam ejusdem esse speciei.—*Linnæi Amœnitates Acad.* "*Anthropomorpha.*"

THE question of questions for mankind—the problem which underlies all others, and is more deeply interesting than any other—is the ascertainment of the place which Man occupies in nature and of his relations to the universe of things. Whence our race has come; what are the limits of our power over nature, and of nature's power over us; to what goal we are tending; are the problems which present themselves anew and with undiminished interest to every man born into the world. Most of us, shrinking from the difficulties and dangers which beset the seeker after original answers to these riddles, are contented to ignore them altogether, or to smother the investigating spirit under the feather-bed of respected and respectable tradition. But, in every age, one or two restless spirits, blessed with that constructive genius, which can only build on a secure foundation, or cursed with the spirit of mere scepticism, are unable to follow in the well-worn and comfortable track of their forefathers and contemporaries, and unmindful of thorns and stumbling-blocks, strike out into paths of their own. The sceptics end in the infidelity which asserts the problem to be insoluble, or in the atheism which denies the existence of any orderly progress and governance of things: the men of genius propound solutions which grow into systems of Theology or of Philosophy, or veiled in musical language which suggests more than it asserts, take the shape of the Poetry of an epoch.

Each such answer to the great question, invariably asserted by the followers of its propounder, if not by himself, to be complete and final, remains in high authority and esteem, it may be for one century, or it may be for twenty: but, as invariably, Time proves each reply to have been a mere approximation to the truth—tolerable chiefly on account of the ignorance of those by whom it was accepted, and wholly intolerable when tested by the larger knowledge of their successors.

In a well-worn metaphor, a parallel is drawn between the life of man and the metamorphosis of the caterpillar into the butterfly; but the comparison may be more just as well as more novel, if for its former term we take the mental progress of the race. History shows that the human mind, fed by constant accessions of knowledge, periodically grows too large for its theoretical coverings, and bursts them asunder to appear in new habiliments, as the feeding and growing grub, at intervals, casts its too narrow skin and assumes another, itself but temporary. Truly the imago state of Man seems to be terribly distant, but every moult is a step gained, and of such there have been many.

Since the revival of learning, whereby the Western races of Europe were enabled to enter upon that progress towards true knowledge, which was commenced by the philosophers of Greece, but was almost arrested in subsequent long ages of intellectual stagnation, or, at most, gyration, the human larva has been feeding vigorously, and moulting in proportion. A skin of some dimension was cast in the 16th century, and another towards the end of the 18th, while, within the last fifty years, the extraordinary growth of every department of physical science has spread among us mental food of so nutritious and stimulating a character that a new ecdysis seems imminent. But this is a process not unusually accompanied by many throes and some sickness and debility, or, it may be, by graver disturbances; so that every good citizen must feel bound to facilitate the process, and even if he have nothing but a scalpel to work withal, to ease the cracking integument to the best of his ability.

In this duty lies my excuse for the publication of these essays. For it will be admitted that some knowledge of man's position in the animate world is an indispensable preliminary to the proper understanding of his relations to the universe; and this again resolves itself, in the long run, into an inquiry into the nature and the closeness of the ties which connect him with those singular creatures whose history[1] has been sketched in the preceding pages.

The importance of such an inquiry is indeed intuitively manifest. Brought face to face with these blurred copies of himself, the least thoughtful of men is conscious of a certain shock, due perhaps, not so much to disgust at the aspect of what looks like an insulting caricature, as to the awakening of a sudden and profound mistrust of time-honoured theories and strongly-rooted prejudices regarding his own position in nature, and

[1] It will be understood that, in the preceding Essay, I have selected for notice from the vast mass of papers which have been written upon the man-like Apes, only those which seem to me to be of special moment.

his relations to the under-world of life; while that which remains a dim suspicion for the unthinking, becomes a vast argument, fraught with the deepest consequences, for all who are acquainted with the recent progress of the anatomical and physiological sciences.

I now propose briefly to unfold that argument, and to set forth, in a form intelligible to those who possess no special acquaintance with anatomical science, the chief facts upon which all conclusions respecting the nature and the extent of the bonds which connect man with the brute world must be based: I shall then indicate the one immediate conclusion which, in my judgment, is justified by those facts, and I shall finally discuss the bearing of that conclusion upon the hypotheses which have been entertained respecting the Origin of Man.

The facts to which I would first direct the reader's attention, though ignored by many of the professed instructors of the public mind, are easy of demonstration and are universally agreed to by men of science; while their significance is so great, that whoso has duly pondered over them will, I think, find little to startle him in the other revelations of Biology. I refer to those facts which have been made known by the study of Development.

It is a truth of very wide, if not of universal,

application, that every living creature commences its existence under a form different from, and simpler than, that which it eventually attains.

The oak is a more complex thing than the little rudimentary plant contained in the acorn; the caterpillar is more complex than the egg; the butterfly than the caterpillar; and each of these beings, in passing from its rudimentary to its perfect condition, runs through a series of changes, the sum of which is called its Development. In the higher animals these changes are extremely complicated; but, within the last half century, the labours of such men as Von Baer, Rathke, Reichert, Bischoff, and Remak, have almost completely unravelled them, so that the successive stages of development which are exhibited by a Dog, for example, are now as well known to the embryologist as are the steps of the metamorphosis of the silk-worm moth to the school-boy. It will be useful to consider with attention the nature and the order of the stages of canine development, as an example of the process in the higher animals generally.

The dog, like all animals, save the very lowest (and further inquiries may not improbably remove the apparent exception), commences its existence as an egg: as a body which is, in every sense, as much an egg as that of a hen, but is devoid of that accumulation of nutritive matter which confers upon the bird's egg its exceptional size and domestic utility; and wants the shell, which would not only be useless to an animal incubated within the body of its parent, but would cut it off from access to the source of that nutriment which the young creature requires, but which the minute egg of the mammal does not contain within itself.

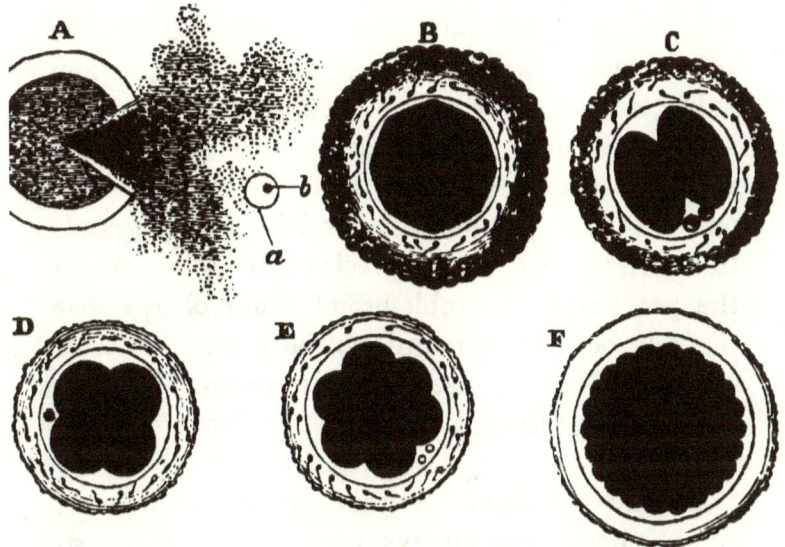

Fig. 13.—A. Egg of the Dog, with the vitelline membrane burst, so as to give exit to the yelk, the germinal vesicle (*a*), and its included spot (*b*). B. C. D. E. F. Successive changes of the yelk indicated in the text. After Bischoff.

The Dog's egg is, in fact, a little spheroidal bag (Fig. 13), formed of a delicate transparent membrane called the *vitelline membrane*, and about $\frac{1}{130}$th to $\frac{1}{120}$th of an inch in diameter. It contains a mass of viscid nutritive matter—the *yelk*—within which is enclosed a second much more delicate spheroidal bag, called the *germinal vesicle* (a). In this, lastly, lies a more solid rounded body, termed the *germinal spot* (b).

The egg, or *Ovum* is originally formed within a gland, from which, in due season, it becomes detached, and passes into the living chamber fitted for its protection and maintenance during the protracted process of gestation. Here, when subjected to the required conditions, this minute and apparently insignificant particle of living matter becomes animated by a new and mysterious activity. The germinal vesicle and spot cease to be discernible (their precise fate being one of the yet unsolved problems of embryology), but the yelk becomes circumferentially indented, as if an invisible knife had been drawn round it, and thus appears divided into two hemispheres (Fig. 13, C).

By the repetition of this process in various planes, these hemispheres become subdivided, so that four segments are produced (D); and these, in like manner, divide and subdivide again, until the whole yelk is converted into a mass of granules, each of which consists of a minute spheroid of yelk-substance, inclosing a central particle, the so-called *nucleus* (F). Nature, by this process, has attained much the same result as that which a human artificer arrives at by his operations in a brick-field. She takes the rough plastic material of the yelk and breaks it up into well-shaped tolerably even-sized masses—handy for building up into any part of the living edifice.

Next, the mass of organic bricks, or *cells* as they are technically called, thus formed, acquires an orderly arrangement, becoming converted into a hollow spheroid with double walls. Then, upon one side of this spheroid, appears a thickening, and, by and bye, in the centre of the area of thickening, a straight shallow groove (Fig. 14, A) marks the central line of the edifice which is to be raised, or, in other words, indicates the position of the middle line of the body of the future dog. The substance bounding the groove on each side next rises up into a fold, the rudiment of the side wall of that long cavity, which will eventually lodge the spinal marrow and the brain; and in the floor of this chamber appears a solid cellular cord, the so-called *notochord*. One end of the enclosed cavity dilates to form the head (Fig. 14, B), the other remains narrow, and eventually becomes the tail; the side walls of the body are fashioned out of the downward continuation of the walls of the groove; and from them, by and bye, grow out little buds which, by degrees, assume the shape of limbs. Watching the fashioning process stage by stage, one is forcibly reminded of the modeller in clay. Every part, every organ, is at first, as it were pinched up rudely, and sketched out in the rough; then shaped more accurately; and only, at last, receives the touches which stamp its final character.

Thus, at length, the young puppy assumes such

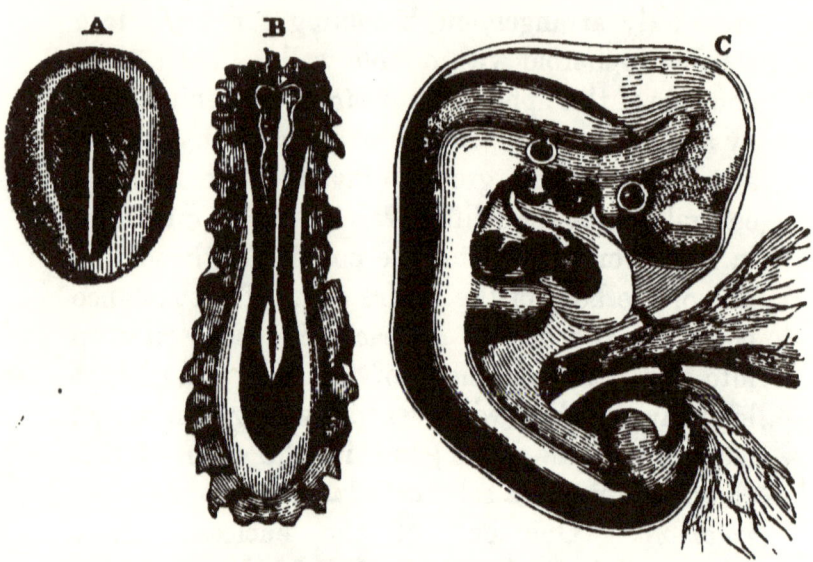

FIG. 14.—A. Earliest rudiment of the Dog. B. Rudiment further advanced, showing the foundations of the head, tail, and vertebral column. C. The very young puppy, with attached ends of the yelk-sac and allantois, and invested in the amnion.

a form as is shown in Fig. 14, C. In this condition it has a disproportionately large head, as dissimilar to that of a dog as the bud-like limbs are unlike his legs.

The remains of the yelk, which have not yet been applied to the nutrition and growth of the young animal, are contained in a sac attached to the rudimentary intestine, and termed the yelk sac, or *umbilical vesicle*. Two membranous bags, intended to subserve respectively the protection and nutrition of the young creature, have been developed from the skin and from the under and hinder surface of the body; the former, the so-called *amnion*, is a sac filled with fluid, which invests the whole body of the embryo, and plays the part of a sort of water-bed for it; the other, termed the *allantois*, grows out, loaded with blood-vessels, from the ventral region, and eventually applying itself to the walls of the cavity, in which the developing organism is contained, enables these vessels to become the channel by which the stream of nutriment, required to supply the wants of the offspring, is furnished to it by the parent.

The structure which is developed by the interlacement of the vessels of the offspring with those of the parent, and by means of which the former is enabled to receive nourishment and to get rid of effete matters, is termed the *Placenta*.

It would be tedious, and it is unnecessary for my present purpose, to trace the process of development further; suffice it to say, that, by a long and gradual series of changes, the rudiment here depicted and described, becomes a puppy, is born, and then, by still slower and less perceptible steps, passes into the adult Dog.

There is not much apparent resemblance between a barn-door Fowl and the Dog who protects the farm-yard. Nevertheless the student of development finds, not only that the chick commences its existence as an egg, primarily identical, in all essential respects, with that of the Dog, but that the yelk of this egg undergoes division—that the primitive groove arises, and that the contiguous parts of the germ are fashioned, by precisely similar methods, into a young chick, which, at one stage of its existence, is so like the nascent Dog, that ordinary inspection would hardly distinguish the two.

The history of the development of any other vertebrate animal, Lizard, Snake, Frog, or Fish, tells the same story. There is always, to begin with, an egg having the same essential structure as that of the Dog:—the yelk of that egg always undergoes division, or *segmentation* as it is often called: the ultimate products of that segmentation constitute the building materials for the body of the young animal; and this is built up round a primitive groove, in the floor of which a notochord is developed. Furthermore, there is a period in which the young of all these animals resemble one another, not merely in outward form, but in all essentials of structure, so closely, that the differences between them are inconsiderable, while, in their subsequent course they diverge more and more widely from one another. And it is a general law, that, the more closely any animals resemble one another in adult structure, the longer and the more intimately do their embryos resemble one another: so that, for example, the embryos of a Snake and of a Lizard remain like one another longer than do those of a Snake and of a Bird; and the embryo of a Dog and of a Cat remain like one another for a far longer period than do those of a Dog and a Bird; or of a Dog and an Opossum; or even than those of a Dog and a Monkey.

Thus the study of development affords a clear test of closeness of structural affinity, and one turns with impatience to inquire what results are yielded by the study of the development of Man. Is he something apart? Does he originate in a totally different way from Dog, Bird, Frog, and Fish, thus justifying those who assert him to have no place in nature and no real affinity with the lower world of animal life? Or does he originate in a similar germ, pass through the same slow and gradually progressive modifications, depend on the same contrivances for protection and nutrition, and finally enter the world by the help of the same mechanism? The reply is not doubtful for a moment, and has not been doubtful any time these thirty years. Without question, the mode of origin and the early stages of the development of man are identical with those of the animals immediately below him in the scale:—without a doubt, in these respects, he is far nearer the Apes, than the Apes are to the Dog.

The Human ovum is about $\frac{1}{125}$th of an inch in diameter, and might be described in the same

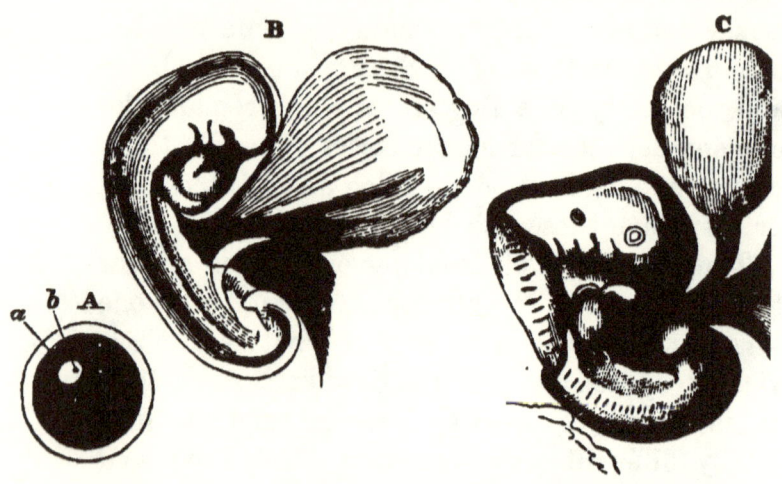

Fig. 15.—A. Human ovum (after Kölliker). *a.* germinal vesicle. *b.* germinal spot. B. A very early condition of Man, with yelk-sac, allantois and amnion (original). C. A more advanced stage (after Kölliker), compare Fig. 14, C.

terms as that of the Dog, so that I need only refer to the figure illustrative (15 A) of its structure. It leaves the organ in which it is formed in a similar fashion and enters the organic chamber prepared for its reception in the same way, the conditions of its development being in all respects the same. It has not yet been possible (and only by some rare chance can it ever be possible) to study the human ovum in so early a developmental stage as that of yelk division, but there is every reason to conclude that the changes it undergoes are identical with those exhibited by the ova of other vertebrated animals; for the formative materials of which the rudimentary human body is composed, in the earliest conditions in which it has been observed, are the same as those of other animals. Some of these earliest stages are figured below and, as will be seen, they are strictly comparable to the very early states of the Dog; the marvellous correspondence between the two which is kept up, even for some time, as development advances, becoming apparent by the simple comparison of the figures with those on page 86.

Indeed, it is very long before the body of the young human being can be readily discriminated from that of the young puppy; but, at a tolerably early period, the two become distinguishable by the different form of their adjuncts, the yelk-sac and the allantois. The former, in the Dog, becomes long and spindle-shaped, while in Man it remains spherical: the latter, in the Dog, attains an extremely large size, and the vascular processes which are developed from it and eventually give rise to the formation of the placenta (taking root, as it were, in the parental organism, so as to draw nourishment therefrom, as the root of a tree extracts it from the soil) are arranged in an encircling zone, while in Man, the allantois remains comparatively small, and its vascular rootlets are eventually restricted to one disk-like spot. Hence, while the placenta of the Dog is like a girdle, that of Man has the cake-like form, indicated by the name of the organ.

But, exactly in those respects in which the developing Man differs from the Dog, he resembles the ape, which, like man, has a spheroidal yelk-sac and a discoidal, sometimes partially lobed, placenta. So that it is only quite in the later stages of development that the young human being presents marked differences from the young ape, while the latter departs as much from the dog in its development, as the man does.

Startling as the last assertion may appear to be, it is demonstrably true, and it alone appears to me sufficient to place beyond all doubt the structural unity of man with the rest of the animal world, and more particularly and closely with the apes.

Thus, identical in the physical processes by which he originates—identical in the early stages of his formation—identical in the mode of his nutrition before and after birth, with the animals which lie immediately below him in the scale—Man, if his adult and perfect structure be compared with theirs, exhibits, as might be expected, a marvellous likeness of organization. He resembles them as they resemble one another—he differs from them as they differ from one another. —And, though these differences and resemblances cannot be weighed and measured, their value may be readily estimated; the scale or standard of judgment, touching that value being afforded and expressed by the system of classification of animals now current among zoologists.

A careful study of the resemblances and differences presented by animals has, in fact, led naturalists to arrange them into groups, or assemblages, all the members of each group presenting a certain amount of definable resemblance, and the number of points of similarity being smaller as the group is larger and *vice versâ*. Thus, all creatures which agree only in presenting the few distinctive marks of animality form the *Kingdom* ANIMALIA. The numerous animals which agree only in possessing the special characters of Vertebrates form one *Sub-kingdom* of this Kingdom. Then the Sub-kingdom VERTEBRATA is subdivided into the five *Classes*, Fishes, Amphibians, Reptiles, Birds, and Mammals, and these into smaller groups called *Orders*; these into *Families* and *Genera*; while the last are finally broken up into the smallest assemblages, which are distinguished by the possession of constant, not-sexual, characters. These ultimate groups are Species.

Every year tends to bring about a greater uniformity of opinion throughout the zoological world as to the limits and characters of these groups, great and small. At present, for example, no one has the least doubt regarding the characters of the classes Mammalia, Aves, or Reptilia; nor does the question arise whether any thoroughly well-known animal should be placed in one class or the other. Again, there is a very general agreement respecting the characters and limits of the orders of Mammals, and as to the animals which are structurally necessitated to take a place in one or another order.

No one doubts, for example, that the Sloth and the Ant-eater, the Kangaroo and the Opossum, the Tiger and the Badger, the Tapir and the Rhinoceros, are respectively members of the same orders. These successive pairs of animals may, and some do, differ from one another immensely, in such matters as the proportions and structure of their limbs; the number of their dorsal and lumbar vertebræ; the adaptation of their frames to climbing, leaping, or running; the number and form of their teeth; and the characters of their skulls and of the contained brain. But, with all these differences, they are so closely connected in all the more important and fundamental characters of their organization, and so distinctly separated by these same characters from other animals, that zoologists find it necessary to group them together as members of one order. And if any new animal were discovered, and were found to present no greater difference from the Kangaroo or from the Opossum, for example, than these animals do from one another, the zoologist would not only be logically compelled to rank it in the same order with these, but he would not think of doing otherwise.

Bearing this obvious course of zoological reasoning in mind, let us endeavour for a moment to disconnect our thinking selves from the mask of humanity; let us imagine ourselves scientific Saturnians, if you will, fairly acquainted with such animals as now inhabit the Earth, and employed in discussing the relations they bear to a new and singular "erect and featherless biped," which some enterprising traveller, overcoming the difficulties of space and gravitation, has brought from that distant planet for our inspection, well preserved, may be, in a cask of rum. We should all, at once, agree upon placing him among the mammalian vertebrates; and his lower jaw, his molars, and his brain, would leave no room for doubting the systematic position of the new genus among those mammals, whose young are nourished during gestation by means of a placenta, or what are called the "placental mammals."

Further, the most superficial study would at once convince us that, among the orders of placental mammals, neither the Whales, nor the hoofed creatures, nor the Sloths and Ant-eaters, nor the carnivorous Cats, Dogs, and Bears, still less the Rodent Rats and Rabbits, or the Insectivorous Moles and Hedgehogs, or the Bats, could claim our *Homo*, as one of themselves.

There would remain then, but one order for comparison, that of the Apes (using that word in its broadest sense), and the question for discussion would narrow itself to this—is Man so different from any of these Apes that he must form an order by himself? Or does he differ less from them than they differ from one another, and hence must take his place in the same order with them?

Being happily free from all real, or imaginary, personal interest in the results of the inquiry thus set afoot, we should proceed to weigh the arguments on one side and on the other, with as much judicial calmness as if the question related to a new Opossum. We should endeavour to ascertain, without seeking either to magnify or diminish them, all the characters by which our new Mammal differed from the Apes; and if we found that these were of less structural value than those which distinguish certain members of the Ape order from others universally admitted to be of the same order, we should undoubtedly place the newly discovered tellurian genus with them.

I now proceed to detail the facts which seem to

me to leave us no choice but to adopt the last-mentioned course.

It is quite certain that the Ape which most nearly approaches man, in the totality of its organisation, is either the Chimpanzee or the Gorilla; and as it makes no practical difference, for the purposes of my present argument, which is selected for comparison, on the one hand, with Man, and on the other hand, with the rest of the Primates,[1] I shall select the latter (so far as its organisation is known)—as a brute now so celebrated in prose and verse, that all must have heard of him, and have formed some conception of his appearance. I shall take up as many of the most important points of difference between man and this remarkable creature, as the space at my disposal will allow me to discuss, and the necessities of the argument demand; and I shall inquire into the value and magnitude of these differences, when placed side by side with those which separate the Gorilla from other animals of the same order.

In the general proportions of the body and limbs there is a remarkable difference between the Gorilla and Man, which at once strikes the

---

[1] We are not at present thoroughly acquainted with the brain of the Gorilla, and therefore, in discussing cerebral characters, I shall take that of the Chimpanzee as my highest term among the Apes.

eye. The Gorilla's brain-case is smaller, its trunk larger, its lower limbs shorter, its upper limbs longer in proportion than those of Man.

I find that the vertebral column of a full-grown Gorilla, in the Museum of the Royal College of Surgeons, measures 27 inches along its anterior curvature, from the upper edge of the atlas, or first vertebra of the neck, to the lower extremity of the sacrum; that the arm, without the hand, is $31\frac{1}{2}$ inches long; that the leg, without the foot, is $26\frac{1}{2}$ inches long; that the hand is $9\frac{3}{4}$ inches long; the foot $11\frac{1}{4}$ inches long.

In other words, taking the length of the spinal column as 100, the arm equals 115, the leg 96, the hand 36, and the foot 41.

In the skeleton of a male Bosjesman, in the same collection, the proportions, by the same measurement, to the spinal column, taken as 100, are—the arm 78, the leg 110, the hand 26, and the foot 32. In a woman of the same race the arm is 83, and the leg 120, the hand and foot remaining the same. In a European skeleton I find the arm to be 80, the leg 117, the hand 26, the foot 35.

Thus the leg is not so different as it looks at first sight, in its proportion to the spine in the Gorilla and in the Man—being very slightly shorter than the spine in the former, and between $\frac{1}{10}$ and $\frac{1}{5}$ longer than the spine in the latter. The foot is longer and the hand much longer in the Gorilla; but the great difference is caused by the arms, which are very much longer than the spine in the Gorilla, very much shorter than the spine in the Man.

The question now arises how are the other Apes related to the Gorilla in these respects—taking the length of the spine, measured in the same way, at 100. In an adult Chimpanzee, the arm is only 96, the leg 90, the hand 43, the foot 39—so that the hand and the leg depart more from the human proportion and the arm less, while the foot is about the same as in the Gorilla.

In the Orang, the arms are very much longer than in the Gorilla (122), while the legs are shorter (88); the foot is longer than the hand (52 and 48), and both are much longer in proportion to the spine.

In the other man-like Apes again, the Gibbons, these proportions are still further altered; the length of the arms being to that of the spinal column as 19 to 11; while the legs are also a third longer than the spinal column, so as to be longer than in Man, instead of shorter. The hand is half as long as the spinal column, and the foot, shorter than the hand, is about $\frac{5}{11}$ths of the length of the spinal column.

Thus *Hylobates* is as much longer in the arms than the Gorilla, as the Gorilla is longer in the arms than Man; while, on the other hand, it is as much longer in the legs than the Man, as the

Man is longer in the legs than the Gorilla, so that it contains within itself the extremest deviations from the average length of both pairs of limbs.[1]

The Mandrill presents a middle condition, the arms and legs being nearly equal in length, and both being shorter than the spinal column; while hand and foot have nearly the same proportions to one another and to the spine, as in Man.

In the Spider monkey (*Ateles*) the leg is longer than the spine, and the arm than the leg; and, finally, in that remarkable Lemurine form, the Indri (*Lichanotus*), the leg is about as long as the spinal column, while the arm is not more than $\frac{11}{18}$ of its length; the hand having rather less and the foot rather more, than one third the length of the spinal column.

These examples might be greatly multiplied, but they suffice to show that, in whatever proportion of its limbs the Gorilla differs from Man, the other Apes depart still more widely from the Gorilla and that, consequently, such differences of proportion can have no ordinal value.

We may next consider the differences presented by the trunk, consisting of the vertebral column, or backbone, and the ribs and pelvis, or bony hip-basin, which are connected with it, in Man and in the Gorilla respectively.

[1] See the figures of the skeletons of four anthropoid apes and of man, drawn to scale, p. 76.

In Man, in consequence partly of the disposition of the articular surfaces of the vertebræ, and largely of the elastic tension of some of the fibrous bands, or ligaments, which connect these vertebræ together, the spinal column, as a whole, has an elegant S-like curvature, being convex forwards in the neck, concave in the back, convex in the loins, or lumbar region, and concave again in the sacral region; an arrangement which gives much elasticity to the whole backbone, and diminishes the jar communicated to the spine, and through it to the head, by locomotion in the erect position.

Furthermore, under ordinary circumstances, Man has seven vertebræ in his neck, which are called *cervical*; twelve succeed these, bearing ribs and forming the upper part of the back, whence they are termed *dorsal*; five lie in the loins, bearing no distinct, or free, ribs, and are called *lumbar*; five, united together into a great bone, excavated in front, solidly wedged in between the hip bones, to form the back of the pelvis, and known by the name of the *sacrum*, succeed these; and finally, three or four little more or less movable bones, so small as to be insignificant, constitute the *coccyx* or rudimentary tail.

In the Gorilla, the vertebral column is similarly divided into cervical, dorsal, lumbar, sacral, and coccygeal vertebræ, and the total number of cervical and dorsal vertebræ, taken together, is the same as in Man; but the development of a pair of ribs to the first lumbar vertebra, which is an exceptional occurrence in Man, is the rule in the Gorilla; and hence, as lumbar are distinguished from dorsal vertebræ only by the presence or absence of free ribs, the seventeen "dorso-lumbar" vertebræ of the Gorilla are divided into thirteen dorsal and four lumbar, while in Man they are twelve dorsal and five lumbar.

Not only, however, does Man occasionally possess thirteen pair of ribs,[1] but the Gorilla sometimes has fourteen pairs, while an Orang-Utan skeleton in the Museum of the Royal College of Surgeons has twelve dorsal and five lumbar vertebræ, as in Man. Cuvier notes the same number in a *Hylobates*. On the other hand, among the lower Apes, many possess twelve dorsal and six or seven lumbar vertebræ; the Douroucouli has fourteen dorsal and eight lumbar, and a Lemur (*Stenops tardigradus*) has fifteen dorsal and nine lumbar vertebræ.

The vertebral column of the Gorilla, as a whole, differs from that of Man in the less marked char-

---

[1] "More than once," says Peter Camper, "have I met with more than six lumbar vertebræ in man. . . . Once I found thirteen ribs and four lumbar vertebræ." Fallopius noted thirteen pair of ribs and only four lumbar vertebræ; and Eustachius once found eleven dorsal vertebræ and six lumbar vertebræ.— *Œuvres de Pierre Camper*, T. 1, p. 42. As Tyson states, his "Pygmie" had thirteen pair of ribs and five lumbar vertebræ. The question of the curves of the spinal column in the Apes requires further investigation.

acter of its curves, especially in the slighter convexity of the lumbar region. Nevertheless, the curves are present, and are quite obvious in young skeletons of the Gorilla and Chimpanzee which have been prepared without removal of the ligaments. In young Orangs similarly preserved on the other hand, the spinal column is either straight, or even concave forwards, throughout the lumbar region.

Whether we take these characters then, or such minor ones as those which are derivable from the proportional length of the spines of the cervical vertebræ, and the like, there is no doubt whatsoever as to the marked difference between Man and the Gorilla; but there is as little, that equally marked differences, of the very same order, obtain between the Gorilla and the lower Apes.

The Pelvis, or bony girdle of the hips, of Man is a strikingly human part of his organisation; the expanded haunch bones affording support for his viscera during his habitually erect posture, and giving space for the attachment of the great muscles which enable him to assume and to preserve that attitude. In these respects the pelvis of the Gorilla differs very considerably from his (Fig. 16). But go no lower than the Gibbon, and see how vastly more he differs from the Gorilla than the latter does from Man, even in this structure. Look at the flat, narrow haunch bones—the

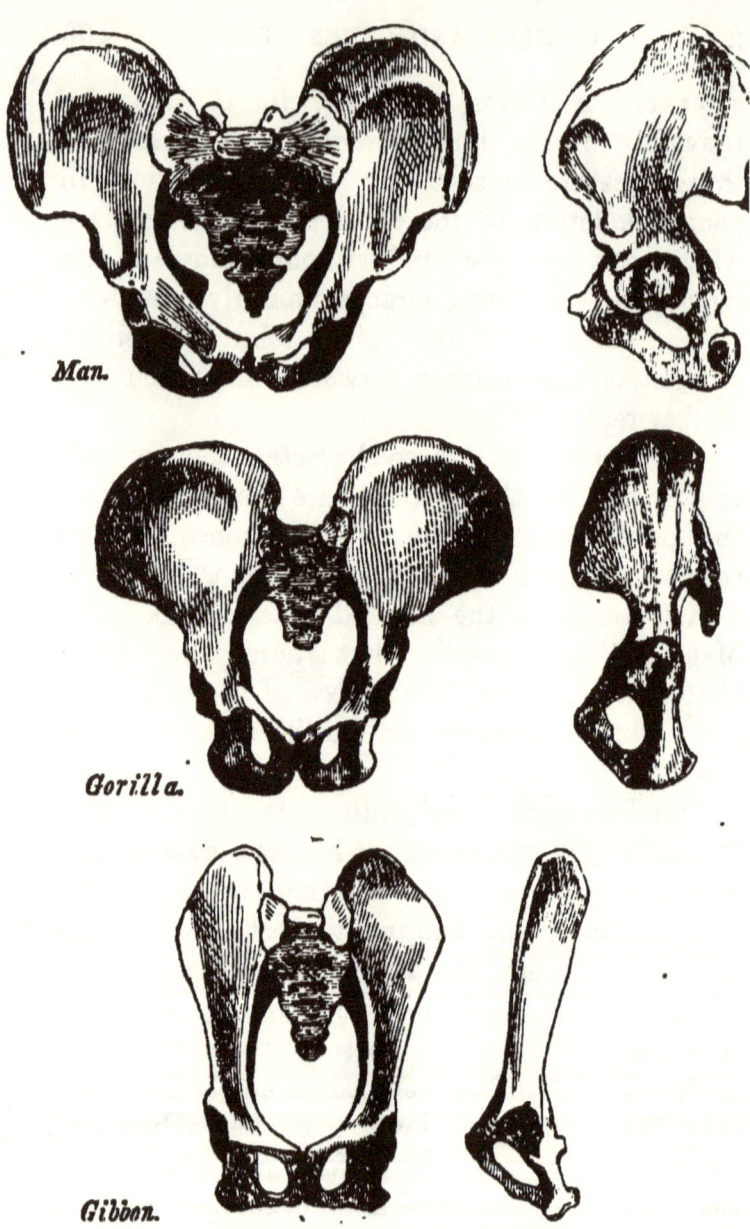

FIG. 16.—Front and side views of the bony pelvis of Man, the Gorilla and Gibbon: reduced from drawings made from nature, of the same absolute length, by Mr. Waterhouse Hawkins.

long and narrow passage—the coarse, outwardly curved, ischiatic prominences on which the Gibbon habitually rests, and which are coated by the so-called "callosities," dense patches of skin, wholly absent in the Gorilla, in the Chimpanzee, and in the Orang, as in Man!

In the lower Monkeys and in the Lemurs the difference becomes more striking still, the pelvis acquiring an altogether quadrupedal character.

But now let us turn to a nobler and more characteristic organ—that by which the human frame seems to be, and indeed is, so strongly distinguished from all others,—I mean the skull. The differences between a Gorilla's skull and a Man's are truly immense (Fig. 17). In the former, the face, formed largely by the massive jaw-bones, predominates over the brain-case, or cranium proper: in the latter, the proportions of the two are reversed. In the Man, the occipital foramen, through which passes the great nervous cord connecting the brain with the nerves of the body, is placed just behind the centre of the base of the skull, which thus becomes evenly balanced in the erect posture; in the Gorilla, it lies in the posterior third of that base. In the Man, the surface of the skull is comparatively smooth, and the supraciliary ridges or brow prominences usually project but little—while, in the Gorilla, vast crests are developed upon the skull, and the brow ridges overhang the cavernous orbits, like great penthouses.

Sections of the skulls, however, show that some of the apparent defects of the Gorilla's cranium arise, in fact, not so much from deficiency of brain-case as from excessive development of the parts of the face. The cranial cavity is not ill-shaped, and the forehead is not truly flattened or very retreating, its really well-formed curve being simply disguised by the mass of bone which is built up against it (Fig. 17).

But the roofs of the orbits rise more obliquely into the cranial cavity, thus diminishing the space for the lower part of the anterior lobes of the brain, and the absolute capacity of the cranium is far less than that of Man. So far as I am aware, no human cranium belonging to an adult man has yet been observed with a less cubical capacity than 62 cubic inches, the smallest cranium observed in any race of men by Morton, measuring 63 cubic inches; while, on the other hand, the most capacious Gorilla skull yet measured has a content of not more than $34\frac{1}{2}$ cubic inches. Let us assume, for simplicity's sake, that the lowest Man's skull has twice the capacity of that of the highest Gorilla.[1]

[1] It has been affirmed that Hindoo crania sometimes contain as little as 27 ounces of water, which would give a capacity of about 46 cubic inches. The minimum capacity which I have assumed above, however, is based upon the valuable tables published by Professor R. Wagner in his *Vorstudien zu einer wissenschaftlichen Morphologie und Physiologie des menschlichen Gehrins*. As the result of the careful weighing of more than 900 human brains, Professor Wagner states that one-half

No doubt, this is a very striking difference, but it loses much of its apparent systematic value, when viewed by the light of certain other equally indubitable facts respecting cranial capacities.

The first of these is, that the difference in the volume of the cranial cavity of different races of mankind is far greater, absolutely, than that between the lowest Man and the highest Ape, while, relatively, it is about the same. For the largest human skull measured by Morton contained 114 cubic inches, that is to say, had very nearly double the capacity of the smallest; while its absolute preponderance, of 52 cubic inches—is far greater than that by which the lowest adult

weighed between 1200 and 1400 grammes, and that about two-ninths, consisting for the most part of male brains, exceed 1400 grammes. The lightest brain of an adult male, with sound mental faculties, recorded by Wagner, weighed 1020 grammes. As a gramme equals 15·4 grains, and a cubic inch of water contains 252·4 grains, this is equivalent to 62 cubic inches of water; so that as brain is heavier than water, we are perfectly safe against erring on the side of diminution in taking this as the smallest capacity of any adult male human brain. The only adult male brain, weighing as little as 970 grammes, is that of an idiot; but the brain of an adult woman, against the soundness of whose faculties nothing appears, weighed as little as 907 grammes (55·3 cubic inches of water); and Reid gives an adult female brain of still smaller capacity. The heaviest brain (1872 grammes, or about 115 cubic inches) was, however, that of a woman; next to it comes the brain of Cuvier (1861 grammes), then Byron (1807 grammes), and then an insane person (1783 grammes). The lightest adult brain recorded (720 grammes) was that of an idiotic female. The brains of five children, four years old, weighed between 1275 and 992 grammes. So that it may be safely said, that an average European child of four years old has a brain twice as large as that of an adult Gorilla.

male human cranium surpasses the largest of the Gorillas (62 — 34½ = 27½). Secondly, the adult crania of Gorillas which have as yet been measured differ among themselves by nearly one-third, the maximum capacity being 34·5 cubic inches, the minimum 24 cubic inches; and, thirdly, after making all due allowance for difference of size, the cranial capacities of some of the lower Apes fall nearly as much, relatively, below those of the higher Apes as the latter fall below Man.

Thus, even in the important matter of cranial capacity, Men differ more widely from one another than they do from the Apes; while the lowest Apes differ as much, in proportion, from the highest, as the latter does from Man. The last proposition is still better illustrated by the study of the modifications which other parts of the cranium undergo in the Simian series.

It is the large proportional size of the facial bones and the great projection of the jaws which confers upon the Gorilla's skull its small facial angle and brutal character.

But if we consider the proportional size of the facial bones to the skull proper only, the little *Chrysothrix* (Fig. 17) differs very widely from the Gorilla, and, in the same way, as Man does; while the Baboons (*Cynocephalus*, Fig. 17) exaggerate the gross proportions of the muzzle of the great Anthropoid, so that its visage looks mild and human by comparison with theirs. The difference

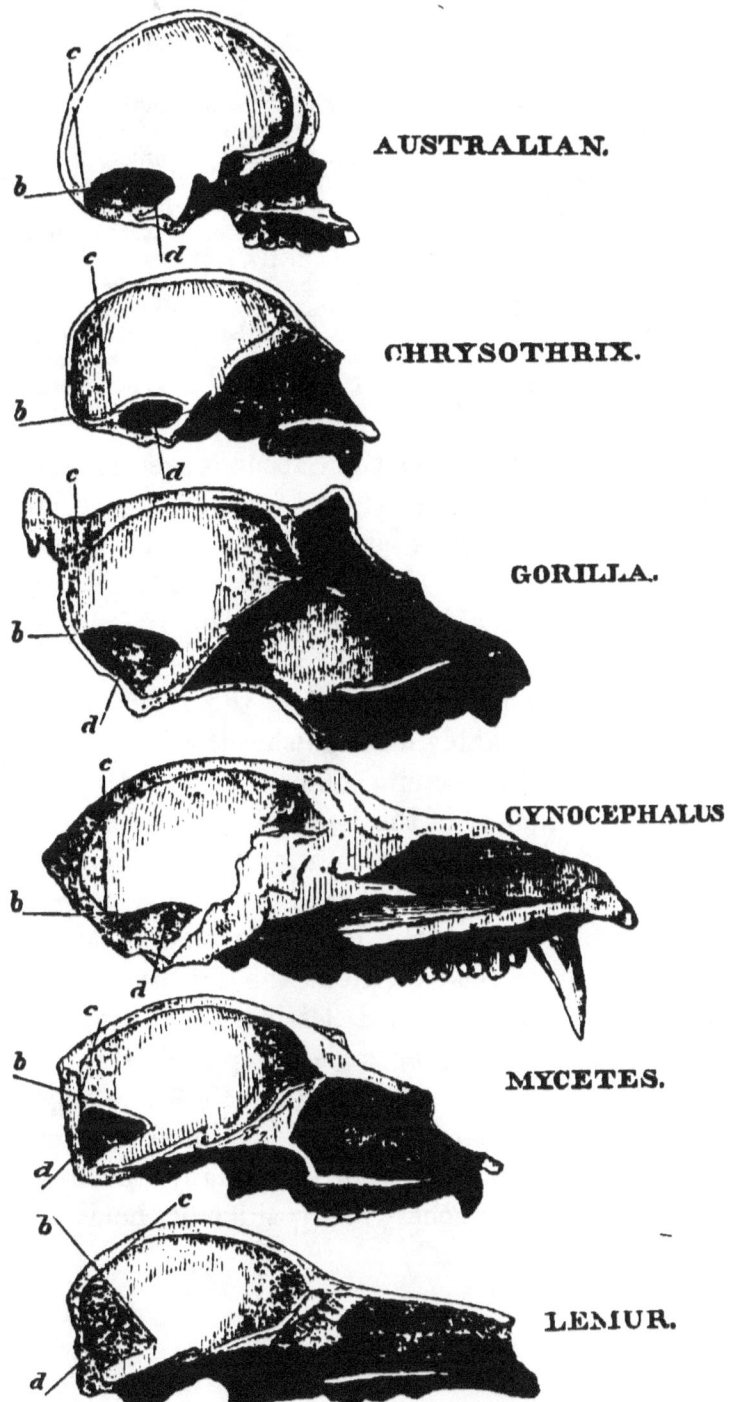

Fig. 17.—Sections of the skulls of Man and various Apes,

drawn so as to give the cerebral cavity the same length in each case, thereby displaying the varying proportions of the facial bones. The line *b* indicates the plane of the tentorium, which separates the cerebrum from the cerebellum; *d*, the axis of the occipital outlet of the skull. The extent of cerebral cavity behind *c*, which is a perpendicular erected on *b* at the point where the tentorium is attached posteriorly, indicates the degree to which the cerebrum overlaps the cerebellum—the space occupied by which is roughly indicated by the dark shading. In comparing these diagrams, it must be recollected, that figures on so small a scale as these simply exemplify the statements in the text, the proof of which is to be found in the objects themselves.

between the Gorilla and the Baboon is even greater than it appears at first sight; for the great facial mass of the former is largely due to a downward development of the jaws; an essentially human character, superadded upon that almost purely forward, essentially brutal, development of the same parts which characterises the Baboon, and yet more remarkably distinguishes the Lemur.

Similarly, the occipital foramen of *Mycetes* (Fig. 17), and still more of the Lemurs, is situated completely in the posterior face of the skull, or as much further back than that of the Gorilla, as that of the Gorilla is further back than that of Man; while, as if to render patent the futility of the attempt to base any broad classificatory distinction on such a character, the same group of Platyrhine, or American monkeys, to which the *Mycetes* belongs, contains the *Chrysothrix*, whose occipital foramen is situated far more forward than in any other ape, and nearly approaches the position it holds in Man.

Again, the Orang's skull is as devoid of excessively developed supraciliary prominences as a Man's, though some varieties exhibit great crests elsewhere (See p. 25); and in some of the Cebine apes and in the *Chrysothrix*, the cranium is as smooth and rounded as that of Man himself.

What is true of these leading characteristics of the skull, holds good, as may be imagined, of all minor features; so that for every constant difference between the Gorilla's skull and the Man's, a similar constant difference of the same order (that is to say, consisting in excess or defect of the same quality) may be found between the Gorilla's skull and that of some other ape. So that, for the skull, no less than for the skeleton in general, the proposition holds good, that the differences between Man and the Gorilla are of smaller value than those between the Gorilla and some other Apes.

In connection with the skull, I may speak of the teeth—organs which have a peculiar classificatory value, and whose resemblances and differences of number, form, and succession, taken as a whole, are usually regarded as more trustworthy indicators of affinity than any others.

Man is provided with two sets of teeth—milk teeth and permanent teeth. The former consist of four incisors, or cutting teeth; two canines, or eye-teeth; and four molars or grinders, in each jaw, making twenty in all. The latter (Fig. 18) comprise four incisors, two canines, four small grinders, called premolars or false molars, and six large grinders, or true molars in each jaw—making thirty-two in all. The internal incisors are larger than the external pair, in the upper jaw, smaller than the external pair, in the lower jaw. The crowns of the upper molars exhibit four cusps, or blunt-pointed elevations, and a ridge crosses the crown obliquely, from the inner, anterior cusp to the outer, posterior cusp (Fig. 18 m^2). The anterior lower molars have five cusps, three external and two internal. The premolars have two cusps, one internal and one external, of which the outer is the higher.

In all these respects the dentition of the Gorilla may be described in the same terms as that of Man; but in other matters it exhibits many and important differences (Fig. 18).

Thus the teeth of man constitute a regular and even series—without any break and without any marked projection of one tooth above the level of the rest; a peculiarity which, as Cuvier long ago showed, is shared by no other mammal save one— as different a creature from man as can well be imagined—namely, the long extinct *Anoplotherium*. The teeth of the Gorilla, on the contrary, exhibit a break, or interval, termed the *diastema*, in both jaws: in front of the eye-tooth, or between it and the outer incisor, in the upper jaw; behind the eye-tooth, or between it and the front false molar, in the

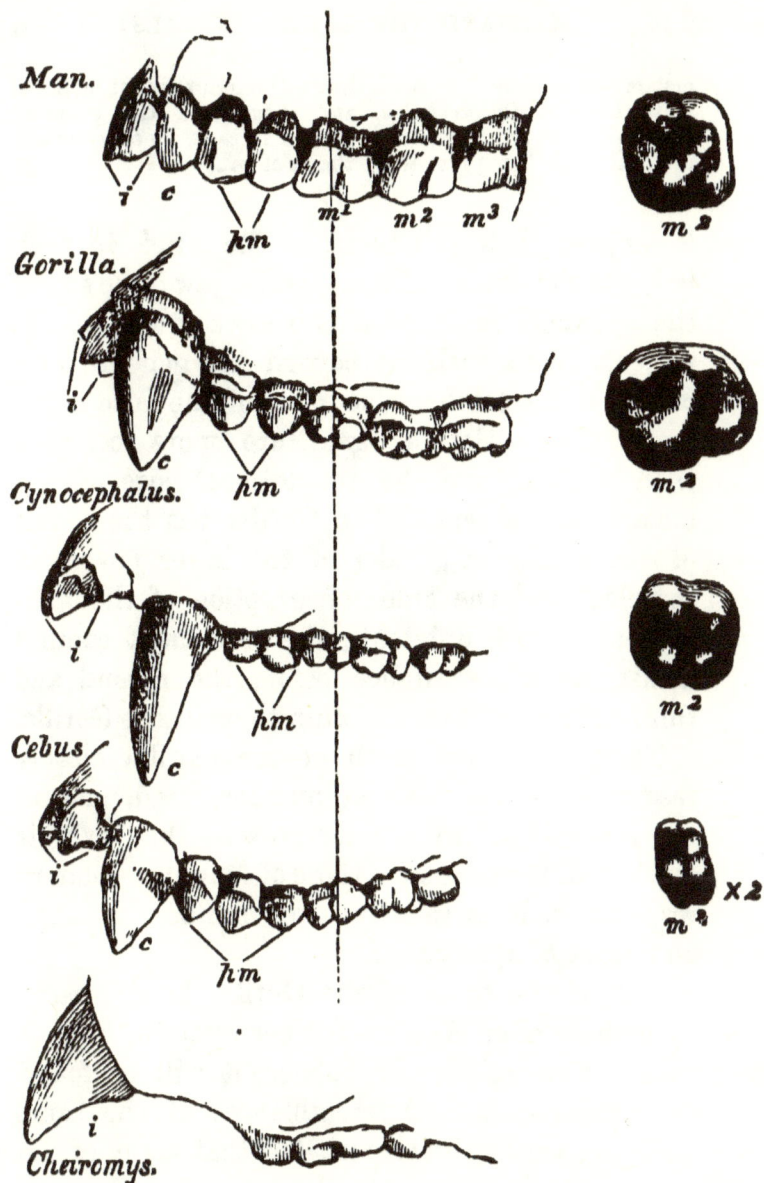

FIG. 18.—Lateral views, of the same length, of the upper jaws of various Primates. *i*, incisors; *c*, canines; *pm*, pre-

molars; *m*, molars. A line is drawn through the first molar of Man, *Gorilla*, *Cynocephalus*, and *Cebus*, and the grinding surface of the second molar is shown in each, its anterior and internal angle being just above the *m* of m^2.

lower jaw. Into this break in the series, in each jaw, fits the canine of the opposite jaw; the size of the eye-tooth in the Gorilla being so great that it projects, like a tusk, far beyond the general level of the other teeth. The roots of the false molar teeth of the Gorilla, again, are more complex than in Man, and the proportional size of the molars is different. The Gorilla has the crown of the hindmost grinder of the lower jaw more complex, and the order of eruption of the permanent teeth is different; the permanent canines making their appearance before the second and third molars in Man, and after them in the Gorilla.

Thus, while the teeth of the Gorilla closely resemble those of Man in number, kind, and in the general pattern of their crowns, they exhibit marked differences from those of Man in secondary respects, such as relative size, number of fangs, and order of appearance.

But, if the teeth of the Gorilla be compared with those of an Ape, no further removed from it than a *Cynocephalus*, or Baboon, it will be found that differences and resemblances of the same order are easily observable; but that many of the points in which the Gorilla resembles Man are those in which it differs from the Baboon; while various respects in which it differs from Man are exaggerated in the *Cynocephalus*. The number and the nature of the teeth remain the same in the Baboon as in the Gorilla and in Man. But the pattern of the Baboon's upper molars is quite different from that described above (Fig. 18), the canines are proportionally longer and more knife-like; the anterior premolar in the lower jaw is specially modified; the posterior molar of the lower jaw is still larger and more complex than in the Gorilla.

Passing from the old-world Apes to those of the new world, we meet with a change of much greater importance than any of these. In such a genus as *Cebus*, for example (Fig. 18), it will be found that while in some secondary points, such as the projection of the canines and the diastema, the resemblance to the great ape is preserved; in other and most important respects, the dentition is extremely different. Instead of 20 teeth in the milk set, there are 24: instead of 32 teeth in the permanent set, there are 36, the false molars being increased from eight to twelve. And in form, the crowns of the molars are very unlike those of the Gorilla, and differ far more widely from the human pattern.

The Marmosets, on the other hand, exhibit the same number of teeth as Man and the Gorilla; but, notwithstanding this, their dentition is very different, for they have four more false molars, like the other American monkeys—but as they have four fewer true molars, the total remains the same. And passing from the American apes to the Lemurs, the dentition becomes still more completely and essentially different from that of the Gorilla. The incisors begin to vary both in number and in form. The molars acquire, more and more, a many-pointed, insectivorous character, and in one Genus, the Aye-Aye (*Cheiromys*), the canines disappear, and the teeth completely simulate those of a Rodent (Fig. 18).

Hence it is obvious that, greatly as the dentition of the highest Ape differs from that of Man, it differs far more widely from that of the lower and lowest Apes.

Whatever part of the animal fabric—whatever series of muscles, whatever viscera might be selected for comparison—the result would be the same—the lower Apes and the Gorilla would differ more than the Gorilla and the Man. I cannot attempt in this place to follow out all these comparisons in detail, and indeed it is unnecessary I should do so. But certain real, or supposed, structural distinctions between man and the apes remain, upon which so much stress has been laid, that they require careful consideration, in order that the true value may be assigned to those which are real, and the emptiness of those which are fictitious may be exposed. I refer to the characters of the hand, the foot, and the brain.

Man has been defined as the only animal possessed of two hands terminating his fore limbs, and of two feet ending his hind limbs, while it has been said that all the apes possess four hands; and he has been affirmed to differ fundamentally from all the apes in the characters of his brain, which alone, it has been strangely asserted and reasserted, exhibits the structures known to anatomists as the posterior lobe, the posterior cornu of the lateral ventricle, and the hippocampus minor.

That the former proposition should have gained general acceptance is not surprising—indeed, at first sight, appearances are much in its favour: but, as for the second, one can only admire the surpassing courage of its enunciator, seeing that it is an innovation which is not only opposed to generally and justly accepted doctrines, but which is directly negatived by the testimony of all original inquirers, who have specially investigated the matter: and that it neither has been, nor can be, supported by a single anatomical preparation. It would, in fact, be unworthy of serious refutation, except for the general and natural belief that deliberate and reiterated assertions must have some foundation.

Before we can discuss the first point with

advantage we must consider with some attention, and compare together, the structure of the human hand and that of the human foot, so that we may have distinct and clear ideas of what constitutes a hand and what a foot.

The external form of the human hand is familiar enough to every one. It consists of a stout wrist followed by a broad palm, formed of flesh, and tendons, and skin, binding together four bones, and dividing into four long and flexible digits, or fingers, each of which bears on the back of its last joint a broad and flattened nail. The longest cleft between any two digits is rather less than half as long as the hand. From the outer side of the base of the palm a stout digit goes off, having only two joints instead of three; so short, that it only reaches to a little beyond the middle of the first joint of the finger next it; and further remarkable by its great mobility, in consequence of which it can be directed outwards, almost at a right angle to the rest. This digit is called the "*pollex*," or thumb; and, like the others, it bears a flat nail upon the back of its terminal joint. In consequence of the proportions and mobility of the thumb, it is what is termed "opposable"; in other words, its extremity can, with the greatest ease, be brought into contact with the extremities of any of the fingers; a property upon which the possibility of our carrying into effect the conceptions of the mind so largely depends.

The external form of the foot differs widely from that of the hand; and yet, when closely compared, the two present some singular resemblances. Thus the ankle corresponds in a manner with the wrist; the sole with the palm; the toes with the fingers; the great toe with the thumb. But the toes, or digits of the foot, are far shorter in proportion than the digits of the hand, and are less moveable, the want of mobility being most striking in the great toe—which, again, is very much larger in proportion to the other toes than the thumb to the fingers. In considering this point, however, it must not be forgotten that the civilized great toe, confined and cramped from childhood upwards, is seen to a great disadvantage, and that in uncivilized and barefooted people it retains a great amount of mobility, and even some sort of opposability. The Chinese boatmen are said to be able to pull an oar; the artisans of Bengal to weave, and the Carajas to steal fishhooks by its help; though, after all, it must be recollected that the structure of its joints and the arrangement of its bones, necessarily render its prehensile action far less perfect than that of the thumb.

But to gain a precise conception of the resemblances and differences of the hand and foot, and of the distinctive characters of each, we must look below the skin, and compare the bony framework and its motor apparatus in each (Fig. 19).

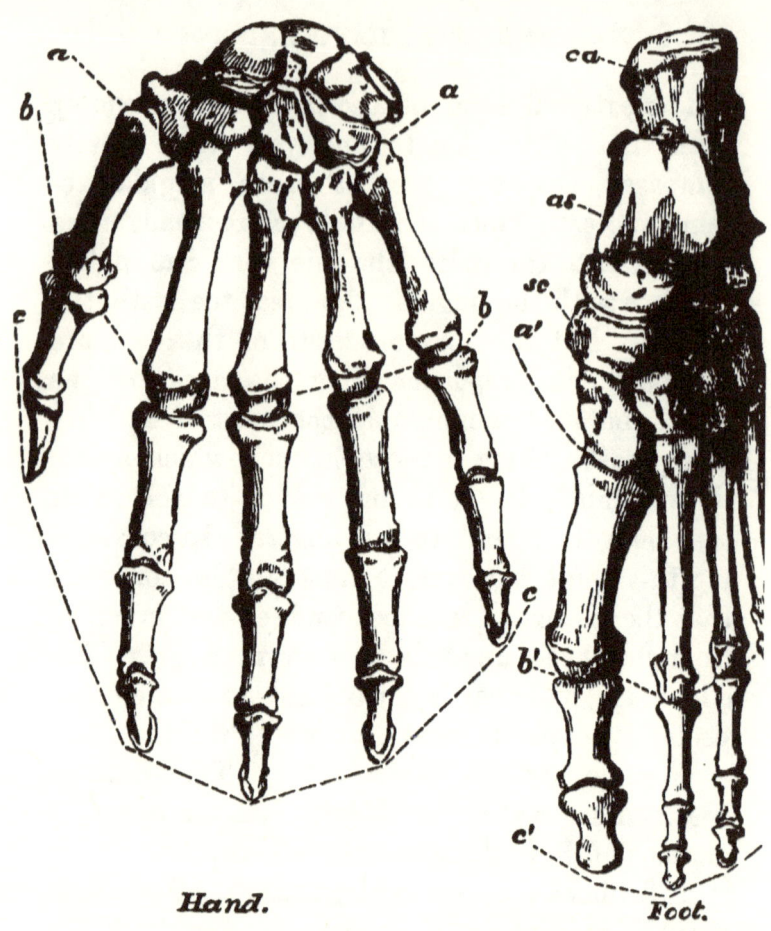

Hand.   *Foot.*

Fig. 19.—The skeleton of the Hand and Foot of Man reduced from Dr. Carter's drawings in Gray's *Anatomy*. The hand is drawn to a larger scale than the foot. The line $a\,a$ in the hand indicates the boundary between the carpus and the metacarpus; $b\,b$ that between the latter and the proximal phalanges; $c\,c$ marks the ends of the distal phalanges. The line $a'\,a'$ in the foot indicates the boundary between the tarsus and metatarsus; $b'\,b'$ marks that between the metatarsus and the proximal phalanges; and $c'\,c'$ bounds the ends of the distal phalanges; *ca*, the calcaneum; *as*, the astragalus; *sc*, the scaphoid bone in the tarsus.

The skeleton of the hand exhibits, in the region which we term the wrist, and which is technically called the *carpus*—two rows of closely fitted polygonal bones, four in each row, which are tolerably equal in size. The bones of the first row with the bones of the forearm, form the wrist joint, and are arranged side by side, no one greatly exceeding or overlapping the rest.

Three of the bones of the second row of the carpus bear the four long bones which support the palm of the hand. The fifth bone of the same character is articulated in a much more free and moveable manner than the others, with its carpal bone, and forms the base of the thumb. These are called *metacarpal* bones, and they carry the *phalanges*, or bones of the digits, of which there are two in the thumb, and three in each of the fingers.

The skeleton of the foot is very like that of the hand in some respects. Thus there are three phalanges in each of the lesser toes, and only two in the great toe, which answers to the thumb. There is a long bone, termed *metatarsal*, answering to the metacarpal, for each digit; and the *tarsus* which corresponds with the carpus, presents four short polygonal bones in a row, which correspond very closely with the four carpal bones of the second row of the hand. In other respects the foot differs very widely from the hand. Thus the great toe is the longest digit but one; and its metatarsal is far less moveably articulated with the tarsus than the metacarpal of the thumb with the carpus. But a far more important distinction lies in the fact that, instead of four more tarsal bones there are only three; and, that these three are not arranged side by side, or in one row. One of them, the *os calcis* or heel bone (*ca*), lies externally, and sends back the large projecting heel; another, the *astragalus* (*as*), rests on this by one face, and by another, forms, with the bones of the leg, the ankle joint; while a third face, directed forwards, is separated from the three inner tarsal bones of the row next the metatarsus by a bone called the *scaphoid* (*sc*).

Thus there is a fundamental difference in the structure of the foot and the hand, observable when the carpus and the tarsus are contrasted: and there are differences of degree noticeable when the proportions and the mobility of the metacarpals and metatarsals, with their respective digits, are compared together.

The same two classes of differences become obvious when the muscles of the hand are compared with those of the foot.

Three principal sets of muscles, called "flexors," bend the fingers and thumb, as in clenching the fist, and three sets,—the extensors—extend them, as in straightening the fingers. These muscles are all "long muscles"; that is to say, the fleshy part of each, lying in and being fixed to the bones of the arm, is, at the other end, continued into tendons, or rounded cords, which pass into the hand, and are ultimately fixed to the bones which are to be moved. Thus, when the fingers are bent, the fleshy parts of the flexors of the fingers, placed in the arm, contract, in virtue of their peculiar endowment as muscles; and pulling the tendinous cords, connecting with their ends, cause them to pull down the bones of the fingers towards the palm.

Not only are the principal flexors of the fingers and of the thumb long muscles, but they remain quite distinct from one another throughout their whole length.

In the foot, there are also three principal flexor muscles of the digits or toes, and three principal extensors; but one extensor and one flexor are short muscles; that is to say, their fleshy parts are not situated in the leg (which corresponds with the arm), but in the back and in the sole of the foot—regions which correspond with the back and the palm of the hand.

Again, the tendons of the long flexor of the toes, and of the long flexor of the great toe, when they reach the sole of the foot, do not remain distinct from one another, as the flexors in the palm of the hand do, but they become united and commingled in a very curious manner—while their united tendons receive an accessory muscle connected with the heel-bone.

But perhaps the most absolutely distinctive character about the muscles of the foot is the existence of what is termed the *peronæus longus*, a long muscle fixed to the outer bone of the leg, and sending its tendon to the outer ankle, behind and below which it passes, and then crosses the foot obliquely to be attached to the base of the great toe. No muscle in the hand exactly corresponds with this, which is eminently a foot muscle.

To resume—the foot of man is distinguished from his hand by the following absolute anatomical differences :—

1. By the arrangement of the tarsal bones.
2. By having a short flexor and a short extensor muscle of the digits.
3. By possessing the muscle termed *peronæus longus*.

And if we desire to ascertain whether the terminal division of a limb, in other Primates, is to be called a foot or a hand, it is by the presence or absence of these characters that we must be guided, and not by the mere proportions and greater or lesser mobility of the great toe, which may vary indefinitely without any fundamental alteration in the structure of the foot.

Keeping these considerations in mind, let us now turn to the limbs of the Gorilla. The terminal division of the fore limb presents no difficulty—bone for bone and muscle for muscle, are found to be arranged essentially as in man, or with such minor differences as are found as varieties in man. The Gorilla's hand is clumsier, heavier, and has a thumb somewhat shorter in proportion than that of man;'but no one has ever doubted it being a true hand.'

At first sight, the termination of the hind limb of the Gorilla looks very hand-like, and as it is still more so in many of the lower apes, it is not wonderful that the appellation " Quadrumana," or four-handed creatures, adopted from the older anatomists[1] by Blumenbach, and unfortunately rendered current by Cuvier, should have gained such wide acceptance as a name for the Simian group. But the most cursory anatomical investigation at once proves that the resemblance of the so-called "hind hand" to a true hand, is only skin deep, and that, in all essential respects, the hind limb of the Gorilla is as truly terminated

---

[1] In speaking of the foot of his "Pygmie," Tyson remarks, p. 13:—
"But this part in the formation and in its function too, being liker a Hand than a Foot: for the distinguishing this sort of animals from others, I have thought whether it might not be reckoned and called rather Quadru-manus than Quadrupes, *i.e.* a four-handed rather than a four-footed animal."

As this passage was published in 1699, M. I. G. St. Hilaire is clearly in error in ascribing the invention of the term "quadrumanous" to Buffon, though "bimanous" may belong to him. Tyson uses "Quadrumanus" in several places, as at p. 91. . . .
"Our *Pygmie* is no Man, nor yet the *common Ape*, but a sort of *Animal* between both; and though a *Biped*, yet of the *Quadrumanus*-kind: though some *Men* too have been observed to use their *Feet* like *Hands* as I have seen several."

by a foot as that of man. The tarsal bones, in all important circumstances of number, disposition, and form, resemble those of man (Fig. 20). The metatarsals and digits, on the other hand, are proportionally longer and more slender, while the great toe is not only proportionally shorter and weaker, but its metatarsal bone is united by a more moveable joint with the tarsus. At the same time, the foot is set more obliquely upon the leg than in man.

As to the muscles, there is a short flexor, a short extensor, and a *peronæus longus*, while the tendons of the long flexors of the great toe and of the other toes are united together and with an accessory fleshy bundle.

The hind limb of the Gorilla, therefore, ends in a true foot, with a very moveable great toe. It is a prehensile foot, indeed, but is in no sense a hand; it is a foot which differs from that of man not in any fundamental character, but in mere proportions, in the degree of mobility, and in the secondary arrangement of its parts.

It must not be supposed, however, because I speak of these differences as not fundamental, that I wish to underrate their value. They are important enough in their way, the structure of the foot being in strict correlation with that of the rest of the organism in each case. Nor can it be doubted that the greater division of physiological labour in Man, so that the function of support is thrown wholly on the leg and foot, is an advance in organization of very great moment to him; but, after all, regarded anatomically, the resemblances between the foot of Man and the foot of the Gorilla are far more striking and important than the differences.

I have dwelt upon this point at length, because it is one regarding which much delusion prevails; but I might have passed it over without detriment to my argument, which only requires me to show that, be the differences between the hand and foot of Man and those of the Gorilla what they may—the differences between those of the Gorilla, and those of the lower Apes are much greater.

It is not necessary to descend lower in the scale than the Orang for conclusive evidence on this head.

The thumb of the Orang differs more from that of the Gorilla than the thumb of the Gorilla differs from that of Man, not only by its shortness, but by the absence of any special long flexor muscle. The carpus of the Orang, like that of most lower apes, contains nine bones, while in the Gorilla, as in Man and the Chimpanzee, there are only eight.

The Orang's foot (Fig. 20) is still more aberrant; its very long toes and short tarsus, short great toe, short and raised heel, great obliquity of articulation with the leg, and absence of a long flexor tendon to the great toe, separating it far more widely from the foot of the Gorilla than the latter is separated from that of Man.

But, in some of the lower apes, the hand and

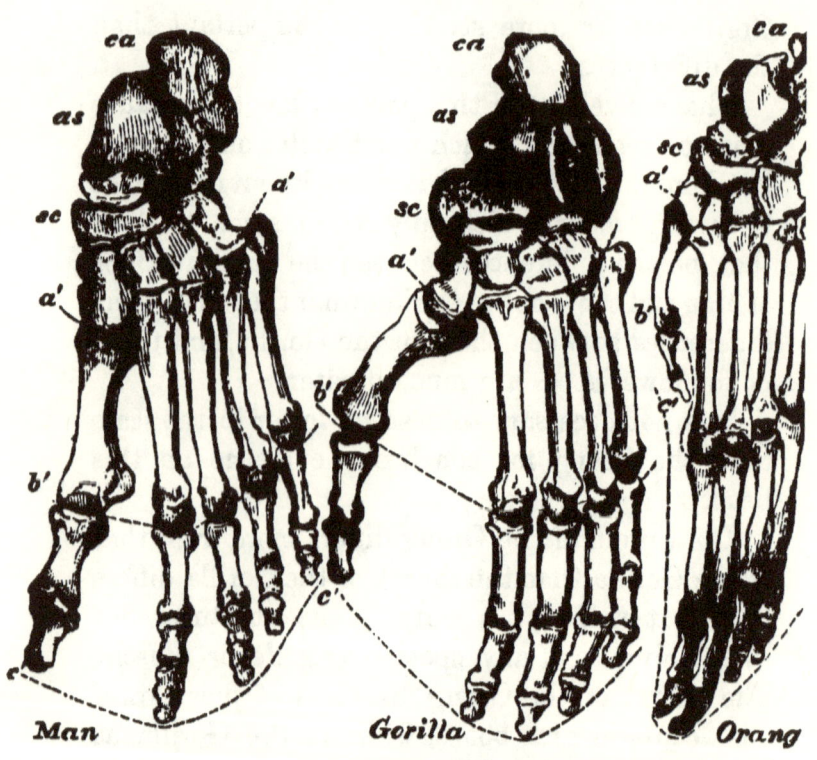

Fig. 20.—Foot of Man, Gorilla, and Orang-Utan of the same absolute length, to show the differences in proportion of each. Letters as in Fig. 19. Reduced from original drawings by Mr. Waterhouse Hawkins.

foot diverge still more from those of the Gorilla, than they do in the Orang. The thumb ceases to be opposable in the American monkeys; is reduced to a mere rudiment covered by the skin in the Spider Monkey; and is directed forwards and armed with a curved claw like the other digits, in the Marmosets—so that, in all these cases, there can be no doubt but that the hand is more different from that of the Gorilla than the Gorilla's hand is from Man's.

And as to the foot, the great toe of the Marmoset is still more insignificant in proportion than that of the Orang—while in the Lemurs it is very large, and as completely thumb-like and opposable as in the Gorilla—but in these animals the second toe is often irregularly modified, and in some species the two principal bones of the tarsus, the *astragalus* and the *os calcis*, are so immensely elongated as to render the foot, so far, totally unlike that of any other mammal.

So with regard to the muscles. The short flexor of the toes of the Gorilla differs from that of Man by the circumstance that one slip of the muscle is attached, not to the heel bone, but to the tendons of the long flexors. The lower Apes depart from the Gorilla by an exaggeration of the same character, two, three, or more, slips becoming fixed to the long flexor tendons—or by a multiplication of the slips.—Again, the Gorilla differs slightly from Man in the mode of interlacing of the long flexor tendons: and the lower apes differ from the Gorilla in exhibiting yet other, sometimes very complex, arrangements of the same parts, and occasionally in the absence of the accessory fleshy bundle.

Throughout all these modifications it must be recollected that the foot loses no one of its essential characters. Every Monkey and Lemur exhibits the characteristic arrangement of tarsal bones, possesses a short flexor and short extensor muscle, and a *peronæus longus*. Varied as the proportions and appearance of the organ may be, the terminal division of the hind limb remains, in plan and principle of construction, a foot, and never, in those respects, can be confounded with a hand.

Hardly any part of the bodily frame, then, could be found better calculated to illustrate the truth that the structural differences between Man and the highest Ape are of less value than those between the highest and the lower Apes, than the hand or the foot; and yet, perhaps, there is one organ the study of which enforces the same conclusion in a still more striking manner—and that is the Brain.

But before entering upon the precise question of the amount of difference between the Ape's brain and that of Man, it is necessary that we should clearly understand what constitutes a great, and what a small difference in cerebral structure; and we shall be best enabled to do this by a brief study of the chief modifications which the brain exhibits in the series of vertebrate animals.

The brain of a fish is very small, compared with the spinal cord into which it is continued, and with the nerves which come off from it: of the segments of which it is composed—the olfactory lobes, the cerebral hemispheres, and the succeeding divisions—no one predominates so much over the rest as to obscure or cover them; and the so-called optic lobes are, frequently, the largest masses of all. In Reptiles, the mass of the brain, relatively to the spinal cord, increases and the cerebral hemispheres begin to predominate over the other parts; while in Birds this predominance is still more marked. The brain of the lowest Mammals, such as the duck-billed Platypus and the Opossums and Kangaroos, exhibits a still more definite advance in the same direction. The cerebral hemispheres have now so much increased in size as, more or less, to hide the representatives of the optic lobes, which remain comparatively small, so that the brain of a Marsupial is extremely different from that of a Bird, Reptile, or Fish. A step higher in the scale, among the placental Mammals, the structure of the brain acquires a vast modification—not that it appears much altered externally, in a Rat or in a Rabbit, from what it is in a Marsupial—nor that the proportions of its parts are much changed, but an apparently new structure is found between the cerebral hemispheres, connecting them together, as what is called the "great commissure" or "corpus callosum." The subject requires careful re-investigation, but if the currently received statements are correct, the appearance of the "corpus callosum" in the placental mammals is the greatest and most sudden modification exhibited by the brain in the whole series of vertebrated animals—it is the greatest leap anywhere made by Nature in her brain work. For the two halves of the brain being once thus knit together, the progress of cerebral complexity is traceable through a complete series of steps from the lowest Rodent, or Insectivore, to Man; and that complexity consists, chiefly, in the disproportionate development of the cerebral hemispheres and of the cerebellum, but especially of the former, in respect to the other parts of the brain.

In the lower placental mammals, the cerebra. hemispheres leave the proper upper and posterior face of the cerebellum completely visible, when the brain is viewed from above; but, in the higher forms, the hinder part of each hemisphere, separated only by the tentorium (p. 137) from the anterior face of the cerebellum, inclines backwards and downwards, and grows out, as the so-called "posterior lobe," so as at length to overlap and hide the cerebellum. In all Mammals, each cerebral hemisphere contains a cavity which is termed the "ventricle"; and as this ventricle is prolonged, on the one hand, forwards, and on the other downwards, into the substance of the hemisphere, it is said to have two horns or "cornua," an "anterior cornu," and a "descending cornu." When the posterior lobe is well developed, a third prolongation of the ventricular cavity extends into it, and is called the "posterior cornu."

In the lower and smaller forms of placental Mammals the surface of the cerebral hemispheres is either smooth or evenly rounded, or exhibits a very few grooves, which are technically termed "sulci," separating ridges or "convolutions" of the substance of the brain; and the smaller species of all orders tend to a similar smoothness of brain. But, in the higher orders, and especially the larger members of these orders, the grooves, or sulci, become extremely numerous, and the intermediate convolutions proportionately more complicated in their meanderings, until, in the Elephant, the Porpoise, the higher Apes, and Man, the cerebral surface appears a perfect labyrinth of tortuous foldings.

Where a posterior lobe exists and presents its customary cavity—the posterior cornu—it commonly happens that a particular sulcus appears upon the inner and under surface of the lobe, parallel with and beneath the floor of the cornu—which is, as it were, arched over the roof of the sulcus. It is as if the groove had been formed by indenting the floor of the posterior horn from without with a blunt instrument, so that the floor should rise as a convex eminence. Now this eminence is what has been termed the "Hippocampus minor;" the "Hippocampus major" being a larger eminence in the floor of the descending cornu. What may be the functional importance of either of these structures we know not.

As if to demonstrate, by a striking example, the impossibility of erecting any cerebral barrier between man and the apes, Nature has provided us, in the latter animals, with an almost complete series of gradations from brains little higher than that of a Rodent, to brains little lower than that of Man. And it is a remarkable circumstance, that though, so far as our present knowledge extends, there *is* one true structural break in the series of forms of Simian brains, this hiatus does not lie between Man and the man-like apes, but between the lower and the lowest Simians; or, in other words, between the old and new world apes and monkeys, and the Lemurs. Every Lemur which has yet been examined, in fact, has its cerebellum partially visible from above, and its posterior lobe, with the contained posterior cornu and hippocampus minor, more or less rudimentary. Every Marmoset, American monkey, old world monkey, Baboon, or Man-like ape, on the contrary, has its cerebellum entirely hidden, posteriorly, by the cerebral lobes, and possesses a large posterior cornu, with a well-developed hippocampus minor.

In many of these creatures, such as the Saimiri (*Chrysothrix*), the cerebral lobes overlap and extend much further behind the cerebellum, in proportion, than they do in man (Fig. 17)—and it is quite certain that, in all, the cerebellum is completely covered behind, by well developed posterior lobes. The fact can be verified by every one who possesses the skull of any old or new world monkey. For, inasmuch as the brain in all mammals completely fills the cranial cavity, it is obvious that a cast of the interior of the skull will reproduce the general form of the brain, at any rate with such minute and, for the present purpose, utterly unimportant differences as may result from the absence of the enveloping membranes of the brain in the dry skull. But if such a cast be made in plaster, and compared with a similar cast of the interior of a human skull, it will be obvious that the cast of the cerebral chamber, representing the cerebrum of the ape, as completely covers over and overlaps the cast of the cerebellar chamber, representing the cerebellum, as it does in the man (Fig. 21). A careless observer, forgetting that a soft structure like the brain loses its proper shape the moment it is taken out of the skull, may indeed mistake the uncovered condition of the cerebellum of an extracted and distorted brain for the natural relations of the parts; but his error must become patent even to himself if he try to replace the brain within the cranial chamber. To

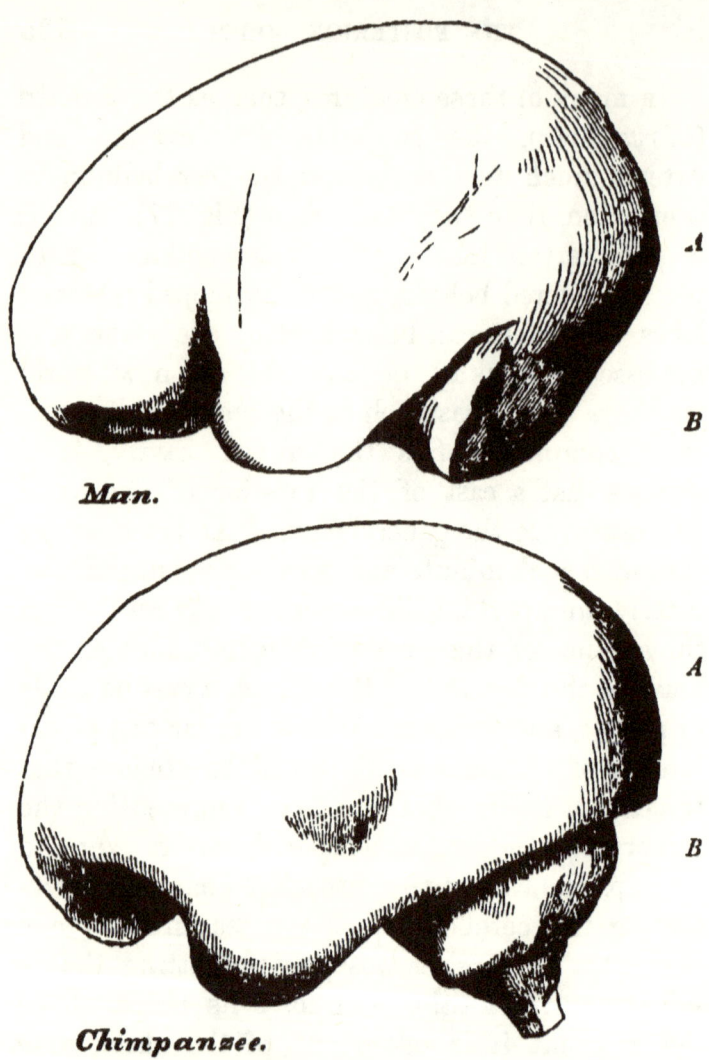

Fig. 21.—Drawings of the internal casts of a Man's and of a Chimpanzee's skull, of the same absolute length, and placed in corresponding positions, *A*. Cerebrum; *B*. Cerebellum. The former drawing is taken from a cast in the Museum of the Royal College of Surgeons, the latter from the photograph of the cast of a Chimpanzee's skull, which illustrates the paper by Mr. Marshall "On the Brain of the Chimpanzee" in the

Natural History Review for July, 1861. The sharper definition of the lower edge of the cast of the cerebral chamber in the Chimpanzee arises from the circumstance that the tentorium remained in that skull and not in the Man's. The cast more accurately represents the brain in the Chimpanzee than in the Man; and the great backward projection of the posterior lobes of the cerebrum of the former, beyond the cerebellum, is conspicuous.

suppose that the cerebellum of an ape is naturally uncovered behind is a miscomprehension comparable only to that of one who should imagine that a man's lungs always occupy but a small portion of the thoracic cavity, because they do so when the chest is opened, and their elasticity is no longer neutralized by the pressure of the air.

And the error is the less excusable, as it must become apparent to every one who examines a section of the skull of any ape above a Lemur, without taking the trouble to make a cast of it. For there is a very marked groove in every such skull, as in the human skull—which indicates the line of attachment of what is termed the *tentorium* —a sort of parchment-like shelf, or partition, which, in the recent state, is interposed between the cerebrum and cerebellum, and prevents the former from pressing upon the latter. (See Fig. 17.)

This groove, therefore, indicates the line of separation between that part of the cranial cavity which contains the cerebrum, and that which contains the cerebellum; and as the brain exactly fills the cavity of the skull, it is obvious that the relations of these two parts of the cranial cavity at once informs us of the relations of their contents. Now in man, in all the old world, and in all the new world Simiæ, with one exception, when the face is directed forwards, this line of attachment of the tentorium, or impression for the lateral sinus, as it is technically called, is nearly horizontal, and the cerebral chamber invariably overlaps or projects behind the cerebellar chamber. In the Howler Monkey or *Mycetes* (see Fig. 17), the line passes obliquely upwards and backwards, and the cerebral overlap is almost nil; while in the Lemurs, as in the lower mammals, the line is much more inclined in the same direction, and the cerebellar chamber projects considerably beyond the cerebral.

When the gravest errors respecting points so easily settled as this question respecting the posterior lobes, can be authoritatively propounded, it is no wonder that matters of observation, of no very complex character, but still requiring a certain amount of care, should have fared worse. Any one who cannot see the posterior lobe in an ape's brain is not likely to give a very valuable opinion respecting the posterior cornu or the hippocampus minor. If a man cannot see a church, it is preposterous to take his opinion about its altar-piece or painted window—so that I do not feel bound to enter upon any discussion of these points, but content myself with assuring the reader that the posterior cornu and the hippocampus minor, have now been seen—usually, at least as well developed as in man, and often better—not only in the Chimpanzee, the Orang, and the Gibbon, but in all the genera of the old world baboons and monkeys, and in most of the new world forms, including the Marmosets.

In fact, all the abundant and trustworthy evidence (consisting of the results of careful investigations directed to the determination of these very questions, by skilled anatomists) which we now possess, leads to the conviction that, so far from the posterior lobe, the posterior cornu, and the hippocampus minor, being structures peculiar to and characteristic of man, as they have been over and over again asserted to be, even after the publication of the clearest demonstration of the reverse, it is precisely these structures which are the most marked cerebral characters common to man with the apes. They are among the most distinctly Simian peculiarities which the human organism exhibits.

As to the convolutions, the brains of the apes exhibit every stage of progress, from the almost smooth brain of the Marmoset, to the Orang and the Chimpanzee, which fall but little below Man. And it is most remarkable that, as soon as all the principal sulci appear, the pattern according to which they are arranged is identical with that of the corresponding sulci of man. The surface of the brain of a monkey exhibits a sort of skeleton map of man's, and in the man-like apes the details become more and more filled in, until it is only in minor characters, such as the greater excavation of the anterior lobes, the constant presence of fissures usually absent in man, and the different disposition and proportions of some convolutions, that the Chimpanzee's or the Orang's brain can be structurally distinguished from Man's.

So far as cerebral structure goes, therefore, it is clear that Man differs less from the Chimpanzee or the Orang, than these do even from the Monkeys, and that the difference between the brains of the Chimpanzee and of Man is almost insignificant, when compared with that between the Chimpanzee brain and that of a Lemur.

It must not be overlooked, however, that there is a very striking difference in absolute mass and weight between the lowest human brain and that of the highest ape—a difference which is all the more remarkable when we recollect that a full-grown Gorilla is probably pretty nearly twice as heavy as a Bosjesman, or as many an European woman. It may be doubted whether a healthy human adult brain ever weighed less than thirty-one or two ounces, or that the heaviest Gorilla brain has exceeded twenty ounces.

This is a very noteworthy circumstance, and doubtless will one day help to furnish an explanation of the great gulf which intervenes between the

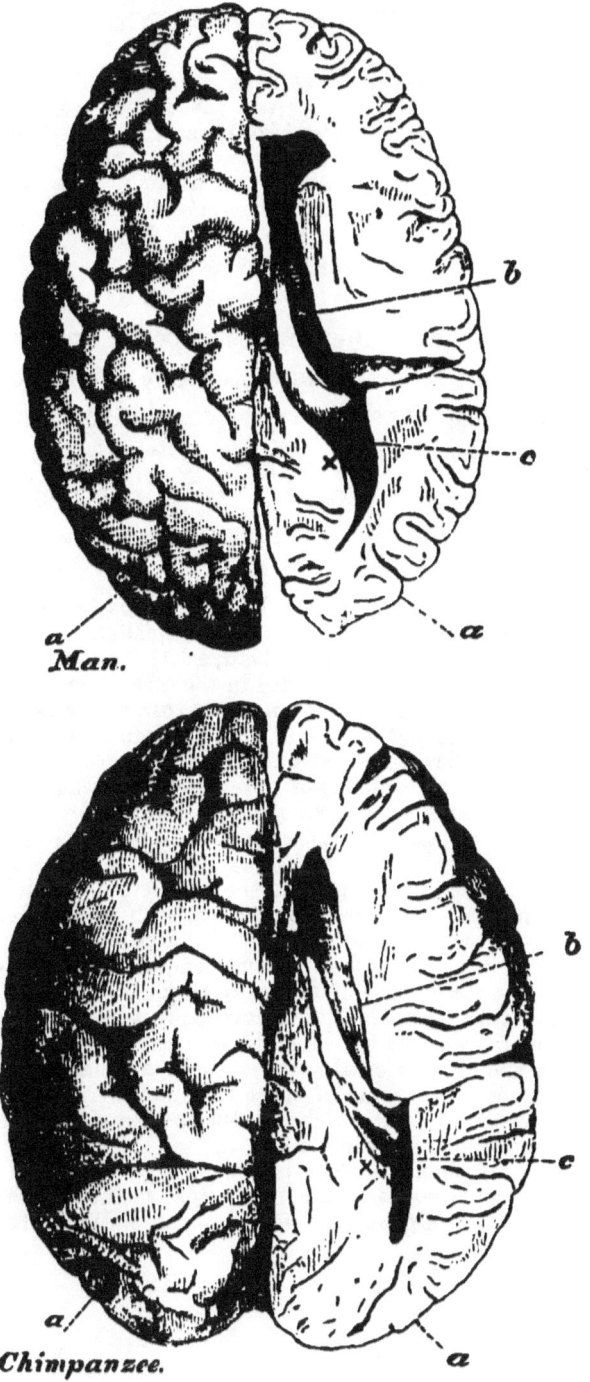

FIG. 22.—Drawings of the cerebral hemispheres of a Man

and of a Chimpanzee of the same length, in order to show the relative proportions of the parts: the former taken from a specimen, which Mr. Flower, Conservator of the Museum of the Royal College of Surgeons, was good enough to dissect for me; the latter, from the photograph of a similarly dissected Chimpanzee's brain, given in Mr. Marshall's paper above referred to. *a*, posterior lobe ; *b*, lateral ventricle ; *c*, posterior cornu ; *x*, the hippocampus minor.

lowest man and the highest ape in intellectual power ;[1] but it has little systematic value, for the simple reason that, as may be concluded from what has been already said respecting cranial capacity, the difference in weight of brain between the highest and the lowest men is far greater, both

[1] I say *help* to furnish: for I by no means believe that it was any original difference of cerebral quality, or quantity, which caused that divergence between the human and the pithecoid stirpes, which has ended in the present enormous gulf between them. It is no doubt perfectly true, in a certain sense, that all difference of function is a result of difference of structure ; or, in other words, of difference in the combination of the primary molecular forces of living substance ; and, starting from this undeniable axiom, objectors occasionally, and with much seeming plausibility, argue that the vast intellectual chasm between the Ape and Man implies a corresponding structural chasm in the organs of the intellectual functions ; so that, it is said, the non-discovery of such vast differences proves, not that they are absent, but that Science is incompetent to detect them. A very little consideration, however, will, I think, show the fallacy of this reasoning. Its validity hangs upon the assumption, that intellectual power depends altogether on the brain—whereas the brain is only one condition out of many on which intellectual manifestations depend ; the others being, chiefly, the organs of the senses and the motor apparatuses, especially those which are concerned in prehension and in the production of articulate speech.

A man born dumb, notwithstanding his great cerebral mass and his inheritance of strong intellectual instincts, would be capable of few higher intellectual manifestations than an

relatively and absolutely, than that between the lowest man and the highest ape. The latter, as has been seen, is represented by, say twelve, ounces of cerebral substance absolutely, or by 32 : 20 relatively; but as the largest recorded human brain weighed between 65 and 66 ounces, the former difference is represented by more than 33 ounces absolutely, or by 65 : 32 relatively. Regarded systematically, the cerebral differences of man and apes, are not of more than generic value; his

Orang or a Chimpanzee, if he were confined to the society of dumb associates. And yet there might not be the slightest discernible difference between his brain and that of a highly intelligent and cultivated person. The dumbness might be the result of a defective structure of the mouth, or of the tongue, or a mere defective innervation of these parts; or it might result from congenital deafness, caused by some minute defect of the internal ear, which only a careful anatomist could discover.

The argument, that because there is an immense difference between a Man's intelligence and an Ape's, therefore, there must be an equally immense difference between their brains, appears to me to be about as well based as the reasoning by which one should endeavour to prove that, because there is a "great gulf" between a watch that keeps accurate time and another that will not go at all, there is therefore a great structural hiatus between the two watches. A hair in the balance-wheel, a little rust on a pinion, a bend in a tooth of the escapement, a something so slight that only the practised eye of the watchmaker can discover it, may be the source of all the difference.

And believing, as I do, with Cuvier, that the possession of articulate speech is the grand distinctive character of man (whether it be absolutely peculiar to him or not), I find it very easy to comprehend, that some equally inconspicuous structural difference may have been the primary cause of the immeasurable and practically infinite divergence of the Human from the Simian Stirps.

Family distinction resting chiefly on his dentition, his pelvis, and his lower limbs.

Thus, whatever system of organs be studied, the comparison of their modifications in the ape series leads to one and the same result—that the structural differences which separate Man from the Gorilla and the Chimpanzee are not so great as those which separate the Gorilla from the lower apes.

But in enunciating this important truth I must guard myself against a form of misunderstanding, which is very prevalent. I find, in fact, that those who endeavour to teach what nature so clearly shows us in this matter, are liable to have their opinions misrepresented and their phraseology garbled, until they seem to say that the structural differences between man and even the highest apes are small and insignificant. Let me take this opportunity then of distinctly asserting, on the contrary, that they are great and significant; that every bone of a Gorilla bears marks by which it might be distinguished from the corresponding bone of a Man; and that, in the present creation, at any rate, no intermediate link bridges over the gap between *Homo* and *Troglodytes*.

It would be no less wrong than absurd to deny the existence of this chasm; but it is at least equally wrong and absurd to exaggerate its magnitude and, resting on the admitted fact of its existence, to refuse to inquire whether it is wide or narrow. Remember, if you will, that there is no existing link between Man and the Gorilla, but do not forget that there is a no less sharp line of demarcation, a no less complete absence of any transitional form, between the Gorilla and the Orang, or the Orang and the Gibbon. I say, not less sharp, though it is somewhat narrower. The structural differences between Man and the Man-like apes certainly justify our regarding him as constituting a family apart from them; though, inasmuch as he differs less from them than they do from other families of the same order, there can be no justification for placing him in a distinct order.

And thus the sagacious foresight of the great lawgiver of systematic zoology, Linnæus, becomes justified, and a century of anatomical research brings us back to his conclusion, that man is a member of the same order (for which the Linnæan term PRIMATES ought to be retained) as the Apes and Lemurs. This order is now divisible into seven families, of about equal systematic value: the first, the ANTHROPINI, contains Man alone; the second, the CATARHINI, embraces the old world apes; the third, the PLATYRHINI, all new world apes, except the Marmosets; the fourth, the ARCTOPITHECINI, contains the Marmosets; the fifth, the LEMURINI, the Lemurs—from which *Cheiromys* should probably be excluded to form a sixth distinct family, the CHEIROMYINI; while the seventh, the GALEOPITHECINI, contains only the flying Lemur *Galeopithecus*,—a strange form which almost touches on the Bats, as the *Cheiromys* puts on a Rodent clothing, and the Lemurs simulate Insectivora.

Perhaps no order of mammals presents us with so extraordinary a series of gradations as this—leading us insensibly from the crown and summit of the animal creation down to creatures, from which there is but a step, as it seems, to the lowest, smallest, and least intelligent of the placental Mammalia. It is as if nature herself had foreseen the arrogance of man, and with Roman severity had provided that his intellect, by its very triumphs, should call into prominence the slaves, admonishing the conqueror that he is but dust.

These are the chief facts, this the immediate conclusion from them to which I adverted in the commencement of this Essay. The facts, I believe, cannot be disputed; and if so, the conclusion appears to me to be inevitable.

But if Man be separated by no greater structural barrier from the brutes than they are from one another—then it seems to follow that if any process of physical causation can be discovered by which the genera and families of ordinary animals have been produced, that process of causation is amply sufficient to account for the origin of Man. In other words, if it could be shown that the Marmosets, for example, have arisen by gradual modification of the ordinary Platyrhini, or that both Marmosets and Platyrhini are modified ramifications of a primitive stock—then, there would be no rational ground for doubting that man might have originated, in the one case, by the gradual modification of a man-like ape; or, in the other case, as a ramification of the same primitive stock as those apes.

At the present moment, but one such process of physical causation has any evidence in its favour; or, in other words, there is but one hypothesis regarding the origin of species of animals in general which has any scientific existence—that propounded by Mr. Darwin. For Lamarck, sagacious as many of his views were, mingled them with so much that was crude and even absurd, as to neutralize the benefit which his originality might have effected, had he been a more sober and cautious thinker; and though I have heard of the announcement of a formula touching "the ordained continuous becoming of organic forms," it is obvious that it is the first duty of a hypothesis to be intelligible, and that a qua-quâ-versal proposition of this kind, which may be read backwards, or forwards, or sideways, with exactly the same amount of signification, does not really exist, though it may seem to do so.

At the present moment, therefore, the question of the relation of man to the lower animals resolves itself, in the end, into the larger question of the tenability, or untenability, of Mr. Darwin's views. But here we enter upon difficult ground, and it behoves us to define our exact position with the greatest care.

It cannot be doubted, I think, that Mr. Darwin has satisfactorily proved that what he terms selection, or selective modification, must occur, and does occur, in nature; and he has also proved to superfluity that such selection is competent to produce forms as distinct, structurally, as some genera even are. If the animated world presented us with none but structural differences, I should have no hesitation in saying that Mr. Darwin had demonstrated the existence of a true physical cause, amply competent to account for the origin of living species, and of man among the rest.

But, in addition to their structural distinctions, the species of animals and plants, or at least a great number of them, exhibit physiological characters—what are known as distinct species, structurally, being for the most part either altogether incompetent to breed one with another; or if they breed, the resulting mule, or hybrid, is unable to perpetuate its race with another hybrid of the same kind.

A true physical cause is, however, admitted to be such only on one condition—that it shall account for all the phenomena which come within the range of its operation. If it is inconsistent with any one phenomenon, it must be rejected; if it fails to explain any one phenomenon, it is so far weak, so far to be suspected; though it may have a perfect right to claim provisional acceptance.

Now, Mr. Darwin's hypothesis is not, so far as I am aware, inconsistent with any known biological fact; on the contrary, if admitted, the facts of Development, of Comparative Anatomy, of Geographical Distribution, and of Palæontology, become connected together, and exhibit a meaning such as they never possessed before; and I, for one, am fully convinced, that if not precisely true, that hypothesis is as near an approximation to the truth as, for example, the Copernican hypothesis was to the true theory of the planetary motions.

But, for all this, our acceptance of the Darwinian hypothesis must be provisional so long as one link in the chain of evidence is wanting; and so long as all the animals and plants certainly produced by selective breeding from a common stock are fertile, and their progeny are fertile with one another, that link will be wanting. For, so long, selective breeding will not be proved to be competent to do all that is required of it to produce natural species.

I have put this conclusion as strongly as possible before the reader, because the last position in which I wish to find myself is that of an advocate for Mr. Darwin's, or any other views; if by an advocate is meant one whose business it is to smooth over real difficulties, and to persuade where he cannot convince.

In justice to Mr. Darwin, however, it must be admitted that the conditions of fertility and sterility are very ill understood, and that every day's advance in knowledge leads us to regard the hiatus in his evidence as of less and less importance, when set against the multitude of facts which harmonize with, or receive an explanation from, his doctrines.

I adopt Mr. Darwin's hypothesis, therefore, subject to the production of proof that physiological species may be produced by selective breeding; just as a physical philosopher may accept the undulatory theory of light, subject to the proof of the existence of the hypothetical ether; or as the chemist adopts the atomic theory, subject to the proof of the existence of atoms; and for exactly the same reasons, namely, that it has an immense amount of primâ facie probability: that it is the only means at present within reach of reducing the chaos of observed facts to order; and lastly, that it is the most powerful instrument of investigation which has been presented to naturalists since the invention of the natural system of classification, and the commencement of the systematic study of embryology.

But even leaving Mr. Darwin's views aside, the whole analogy of natural operations furnishes so complete and crushing an argument against the intervention of any but what are termed secondary causes, in the production of all the phenomena of the universe; that, in view of the intimate relations between Man and the rest of the living world, and between the forces exerted by the latter and all other forces, I can see no excuse for doubting that all are co-ordinated terms of Nature's great progression, from the formless to the formed—from the inorganic to the organic—from blind force to conscious intellect and will.

Science has fulfilled her function when she has ascertained and enunciated truth; and were these pages addressed to men of science only, I should now close this Essay, knowing that my colleagues have learned to respect nothing but evidence, and to believe that their highest duty lies in submitting to it, however it may jar against their inclinations.

But desiring, as I do, to reach the wider circle of the intelligent public, it would be unworthy cowardice were I to ignore the repugnance with which the majority of my readers are likely to meet the conclusions to which the most careful and conscientious study I have been able to give to this matter, has led me.

On all sides I shall hear the cry—" We are men

and women, not a mere better sort of apes, a little longer in the leg, more compact in the foot, and bigger in brain than your brutal Chimpanzees and Gorillas. The power of knowledge—the conscience of good and evil—the pitiful tenderness of human affections, raise us out of all real fellowship with the brutes, however closely they may seem to approximate us."

To this I can only reply that the exclamation would be most just and would have my own entire sympathy, if it were only relevant. But, it is not I who seek to base Man's dignity upon his great toe, or insinuate that we are lost if an Ape has a hippocampus minor. On the contrary, I have done my best to sweep away this vanity. I have endeavoured to show that no absolute structural line of demarcation, wider than that between the animals which immediately succeed us in the scale, can be drawn between the animal world and ourselves; and I may add the expression of my belief that the attempt to draw a psychical distinction is equally futile, and that even the highest faculties of feeling and of intellect begin to germinate in lower forms of life.[1] At the same

---

[1] It is so rare a pleasure for me to find Professor Owen's opinions in entire accordance with my own, that I cannot forbear from quoting a paragraph which appeared in his Essay "On the Characters, &c., of the Class Mammalia," in the *Journal of the Proceedings of the Linnean Society of London* for 1857, but is unaccountably omitted in the "Reade Lecture" delivered before the University of Cambridge two years later

time, no one is more strongly convinced than I am of the vastness of the gulf between civilized man and the brutes; or is more certain that whether *from* them or not, he is assuredly not *of* them. No one is less disposed to think lightly of the present dignity, or desparingly of the future hopes, of the only consciously intelligent denizen of this world.

We are indeed told by those who assume authority in these matters, that the two sets of opinions are incompatible, and that the belief in the unity of origin of man and brutes involves the brutalization and degradation of the former. But is this really so? Could not a sensible child confute by obvious arguments, the shallow rhetoricians who would force this conclusion upon us? Is it, indeed, true, that the Poet, or the Philosopher, or the Artist whose genius is the glory of his age, is degraded from his high estate by the

which is otherwise nearly a reprint of the paper in question. Prof. Owen writes:

"Not being able to appreciate or conceive of the distinction between the psychical phenomena of a Chimpanzee and of a Boschisman or of an Aztec, with arrested brain growth, as being of a nature so essential as to preclude a comparison between them, or as being other than a difference of degree, I cannot shut my eyes to the significance of that all-pervading similitude of structure—every tooth, every bone, strictly homologous— which makes the determination of the difference between *Homo* and *Pithecus* the anatomist's difficulty."

Surely it is a little singular, that the "anatomist," who finds it "difficult" to determine "the difference" between *Homo* and *Pithecus*, should yet range them on anatomical grounds, in distinct sub-classes.

undoubted historical probability, not to say certainty, that he is the direct descendant of some naked and bestial savage, whose intelligence was just sufficient to make him a little more cunning than the Fox, and by so much more dangerous than the Tiger? Or is he bound to howl and grovel on all fours because of the wholly unquestionable fact, that he was once an egg, which no ordinary power of discrimination could distinguish from that of a Dog? Or is the philanthropist, or the saint, to give up his endeavours to lead a noble life, because the simplest study of man's nature reveals, at its foundations, all the selfish passions, and fierce appetites of the merest quadruped? Is mother-love vile because a hen shows it, or fidelity base because dogs possess it?

The common sense of the mass of mankind will answer these questions without a moment's hesitation. Healthy humanity, finding itself hard pressed to escape from real sin and degradation, will leave the brooding over speculative pollution to the cynics and the "righteous overmuch" who, disagreeing in everything else, unite in blind insensibility to the nobleness of the visible world, and in inability to appreciate the grandeur of the place Man occupies therein.

Nay more, thoughtful men, once escaped from the blinding influences of traditional prejudice, will find in the lowly stock whence Man has sprung, the best evidence of the splendour of his capacities; and will discern in his long progress through the Past, a reasonable ground of faith in his attainment of a nobler Future.

They will remember that in comparing civilised man with the animal world, one is as the Alpine traveller, who sees the mountains soaring into the sky and can hardly discern where the deep shadowed crags and roseate peaks end, and where the clouds of heaven begin. Surely the awe-struck voyager may be excused if, at first, he refuses to believe the geologist, who tells him that these glorious masses are, after all, the hardened mud of primeval seas, or the cooled slag of subterranean furnaces—of one substance with the dullest clay, but raised by inward forces to that place of proud and seemingly inaccessible glory.

But the geologist is right; and due reflection on his teachings, instead of diminishing our reverence and our wonder, adds all the force of intellectual sublimity to the mere æsthetic intuition of the uninstructed beholder.

And after passion and prejudice have died away, the same result will attend the teachings of the naturalist respecting that great Alps and Andes of the living world—Man. Our reverence for the nobility of manhood will not be lessened by the knowledge that Man is, in substance and in structure, one with the brutes; for, he alone possesses the marvellous endowment of intelligible and rational speech, whereby, in the secular period of his existence, he has slowly accumulated and organised the experience which is almost wholly lost with the cessation of every individual life in other animals; so that, now, he stands raised upon it as on a mountain top, far above the level of his humble fellows, and transfigured from his grosser nature by reflecting, here and there, a ray from the infinite source of truth.

III

ON SOME FOSSIL REMAINS OF MAN

I HAVE endeavoured to show, in the preceding Essay, that the ANTHROPINI, or Man Family, form a very well-defined group of the Primates, between which and the immediately following Family, the CATARHINI, there is, in the existing world, the same entire absence of any transitional form or connecting link, as between the CATARHINI and PLATYRHINI.

It is a commonly received doctrine, however, that the structural intervals between the various existing modifications of organic beings may be diminished, or even obliterated, if we take into account the long and varied succession of animals and plants which have preceded these now living and which are known to us only by their fossilized remains. How far this doctrine is well based, how far, on the other hand, as our knowledge at present stands, it is an overstatement of the real facts of the case, and an exaggeration of the conclusions fairly deducible from them, are points of grave importance, but into the discussion of which I do not, at present, propose to enter. It is enough that such a view of the relations of extinct to living beings has been propounded, to lead us to inquire, with anxiety, how far the recent discoveries of human remains in a fossil state bear out, or oppose, that view.

I shall confine myself, in discussing this question, to those fragmentary Human skulls from the caves of Engis in the valley of the Meuse, in Belgium, and of the Neanderthal, near Düsseldorf, the geological relations of which have been examined with so much care by Sir Charles Lyell; upon whose high authority I shall take it for granted, that the Engis skull belonged to a contemporary of the Mammoth (*Elephas primigenius*) and of the woolly Rhinoceros (*Rhinoceros tichorhinus*), with the bones of which it was found associated; and that the Neanderthal skull is of great, though uncertain, antiquity. Whatever be the geological age of the latter skull, I conceive it is quite safe (on the ordinary principles of paleontological reasoning) to assume that the former takes us to, at least, the further side of the vague biological limit, which separates the present geological epoch from that which immediately preceded it. And there can be no doubt that the physical geography of Europe has changed wonderfully, since the bones of Men and Mammoths, Hyænas and Rhinoceroses were washed pell-mell into the cave of Engis.

The skull from the cave of Engis was originally

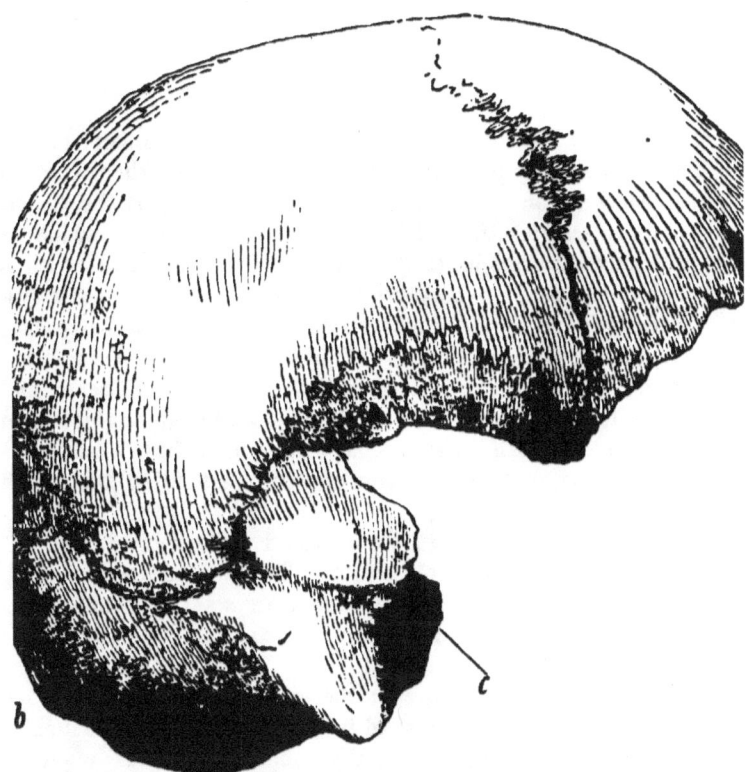

Fig. 23.—The skull from the cave of Engis—viewed from the right side. One half the size of nature. *a* glabella, *b* occipital protuberance (*a* to *b* glabello-occipital line), *c* auditory foramen.

discovered by Professor Schmerling, and was described by him, together with other human remains disinterred at the same time, in his valuable work, "Recherches sur les Ossemens fossiles découverts dans les Cavernes de la Province de Liège," published in 1833 (p. 59, *et seq.*), from which the following paragraphs are extracted, the precise expressions of the author being, as far as possible, preserved.

"In the first place, I must remark that these human remains, which are in my possession, are characterised, like the thousands of bones which I have lately been disinterring, by the extent of the decomposition which they have undergone, which is precisely the same as that of the extinct species: all, with a few exceptions, are broken; some few are rounded, as is frequently found to be the case in fossil remains of other species. The fractures are vertical or oblique; none of them are eroded; their colour does not differ from that of other fossil bones, and varies from whitish yellow to blackish. All are lighter than recent bones, with the exception of those which have a calcareous incrustation, and the cavities of which are filled with such matter.

"The cranium which I have caused to be figured, Plate I, figs. 1, 2, is that of an old person. The sutures are beginning to be effaced: all the facial bones are wanting, and of the temporal bones only a fragment of that of the right side is preserved.

"The face and the base of the cranium had been detached before the skull was deposited in the cave, for we were unable to find those parts, though the whole cavern was regularly searched. The cranium was met with at a depth of a metre and a half [five feet nearly] hidden under an osseous breccia, composed of the remains of small animals, and containing one rhinoceros' tusk, with several teeth of horses and of ruminants. This breccia, which has been spoken of above (p. 31), was a metre [3½ feet about] wide, and rose to the height of a metre and a half above the floor of the cavern, to the walls of which it adhered strongly.

"The earth which contained this human skull exhibited no trace of disturbance: teeth of rhinoceros, horse, hyæna, and bear, surrounded it on all sides.

"The famous Blumenbach [1] has directed attention to the differences presented by the form and the dimensions of human crania of different races. This important work would have assisted us greatly, if the face, a part essential for the determination of race, with more or less accuracy, had not been wanting in our fossil cranium.

"We are convinced that even if the skull had been complete, it would not have been possible to pronounce, with certainty, upon a single specimen; for individual variations are so numerous in the crania of one and the same race, that one cannot, without laying one's self open to large chances of error, draw any inference from a single fragment of a cranium to the general form of the head to which it belonged.

"Nevertheless, in order to neglect no point respecting the form of this fossil skull, we may observe that, from the first, the elongated and narrow form of the forehead attracted our attention.

"In fact, the slight elevation of the frontal, its narrowness, and the form of the orbit, approximate it more nearly to the cranium of an Ethiopian than to that of an European; the elongated form and the produced occiput are also characters which we believe to be observable in our fossil cranium; but to remove all doubt upon that subject I have caused the contours of the cranium of an European and of an Ethiopian to be drawn and the foreheads represented. Plate II, Figs. 1 and 2, and, in the same plate, Figs. 3 and 4, will render the differences easily distinguishable; and a single glance at the figures will be more instructive than a long and wearisome description.

"At whatever conclusion we may arrive as to the origin of the man from whence this fossil skull proceeded, we may express an opinion without exposing ourselves to a fruitless controversy. Each may adopt the hypothesis which seems to him most probable: for my own part, I hold it to be demonstrated that this

---

[1] *Decas Collectionis suæ craniorum diversarum gentium illustrata.*—Gottingæ, 1790-1820.

cranium has belonged to a person of limited intellectual faculties, and we conclude thence that it belonged to a man of a low degree of civilization : a deduction which is borne out by contrasting the capacity of the frontal with that of the occipital region.

"Another cranium of a young individual was discovered in the floor of the cavern beside the tooth of an elephant; the skull was entire when found, but the moment it was lifted it fell into pieces, which I have not, as yet, been able to put together again. But I have represented the bones of the upper jaw, Plate I, Fig. 5. The state of the alveoli and the teeth, shows that the molars had not yet pierced the gum. Detached milk molars and some fragments of a human skull, proceed from this same place. The figure 3 represents a human superior incisor tooth, the size of which is truly remarkable.[1]

"Figure 4 is a fragment of a superior maxillary bone, the molar teeth of which are worn down to the roots.

"I possess two vertebræ, a first and last dorsal.

"A clavicle of the left side (see Plate III, Fig. 1); although it belonged to a young individual, this bone shows that he must have been of great stature.[2]

"Two fragments of the radius, badly preserved, do not indicate that the height of the man, to whom they belonged, exceeded five feet and a half.

"As to the remains of the upper extremities, those which are in my possession consist merely of a fragment of an ulna and of a radius (Plate III, Figs. 5 and 6).

"Figure 2, Plate IV., represents a metacarpal bone, contained in the breccia, of which we have spoken; it was found in the lower part above the cranium: add to this some metacarpal bones, found at very different distances, half-a-dozen metatarsals, three phalanges of the hand, and one of the foot.

---

[1] In a subsequent passage, Schmerling remarks upon the occurrence of an incisor tooth "of enormous size" from the caverns of Engihoul. The tooth figured is somewhat long, but its dimensions do not appear to me to be otherwise remarkable.

[2] The figure of this clavicle measures 5 inches from end to end in a straight line—so that the bone is rather a small than a large one.

> "This is a brief enumeration of the remains of human bones collected in the cavern of Engis, which has preserved for us the remains of three individuals, surrounded by those of the Elephant, of the Rhinoceros, and of Carnivora of species unknown in the present creation."

From the cave of Engihoul, opposite that of Engis, on the right bank of the Meuse, Schmerling obtained the remains of three other individuals of Man, among which were only two fragments of parietal bones, but many bones of the extremities. In one case, a broken fragment of an ulna was soldered to a like fragment of a radius by stalagmite, a condition frequently observed among the bones of the Cave Bear (*Ursus spelæus*), found in the Belgian caverns.

It was in the cavern of Engis that Professor Schmerling found, incrusted with stalagmite and joined to a stone, the pointed bone implement, which he has figured in Fig. 7 of his Plate XXXVI, and worked flints were found by him in all those Belgian caves, which contained an abundance of fossil bones.

A short letter from M. Geoffroy St. Hilaire, published in the "Comptes Rendus" of the Academy of Sciences of Paris, for July 2nd, 1838, speaks of a visit (and apparently a very hasty one) paid to the collection of Professor "Schermidt" (which is presumably a misprint for Schmerling) at Liège. The writer briefly criticises the drawings which illustrate Schmerling's work, and affirms that the "human cranium is a little longer than it is represented" in Schmerling's figure. The only other remark worth quoting is this:—

"The aspect of the human bones differs little from that of the cave bones, with which we are familiar, and of which there is a considerable collection in the same place. With respect to their special forms, compared with those of the varieties of recent human crania, few *certain* conclusions can be put forward ; for much greater differences exist between the different specimens of well-characterized varieties, than between the fossil cranium of Liège and that of one of those varieties selected as a term of comparison."

Geoffroy St. Hilaire's remarks are, it will be observed, little but an echo of the philosophic doubts of the describer and discoverer of the remains. As to the critique upon Schmerling's figures, I find that the side view given by the latter is really about $\frac{3}{10}$ths of an inch shorter than the original, and that the front view is diminished to about the same extent. Otherwise the representation is not, in any way, inaccurate, but corresponds very well with the cast which is in my possession.

A piece of the occipital bone, which Schmerling seems to have missed, has since been fitted on to the rest of the cranium by an accomplished anatomist, Dr. Spring of Liège, under whose direction an excellent plaster cast was made for Sir Charles Lyell. It is upon and from a duplicate of that cast that my own observations and the accompanying figures, the outlines of which are copied from very accurate Camera lucida drawings, by my friend Mr. Busk, reduced to one-half of the natural size, are made.

As Professor Schmerling observes, the base of the skull is destroyed, and the facial bones are entirely absent; but the roof of the cranium, consisting of the frontal, parietal, and the greater part of the occipital bones, as far as the middle of the occipital foramen, is entire, or nearly so. The left temporal bone is wanting. Of the right temporal, the parts in the immediate neighbourhood of the auditory foramen, the mastoid process, and a considerable portion of the squamous element of the temporal are well preserved (Fig. 23).

The lines of fracture which remain between the coadjusted pieces of the skull, and are faithfully displayed in Schmerling's figure, are readily traceable in the cast. The sutures are also discernible, but the complex disposition of their serrations, shown in the figure, is not obvious in the cast. Though the ridges which give attachment to muscles are not excessively prominent, they are well marked, and taken together with the apparently well developed frontal sinuses, and the condition of the sutures, leave no doubt on my mind that the skull is that of an adult, if not middle-aged man.

The extreme length of the skull is 7·7 inches. Its extreme breadth, which corresponds very nearly

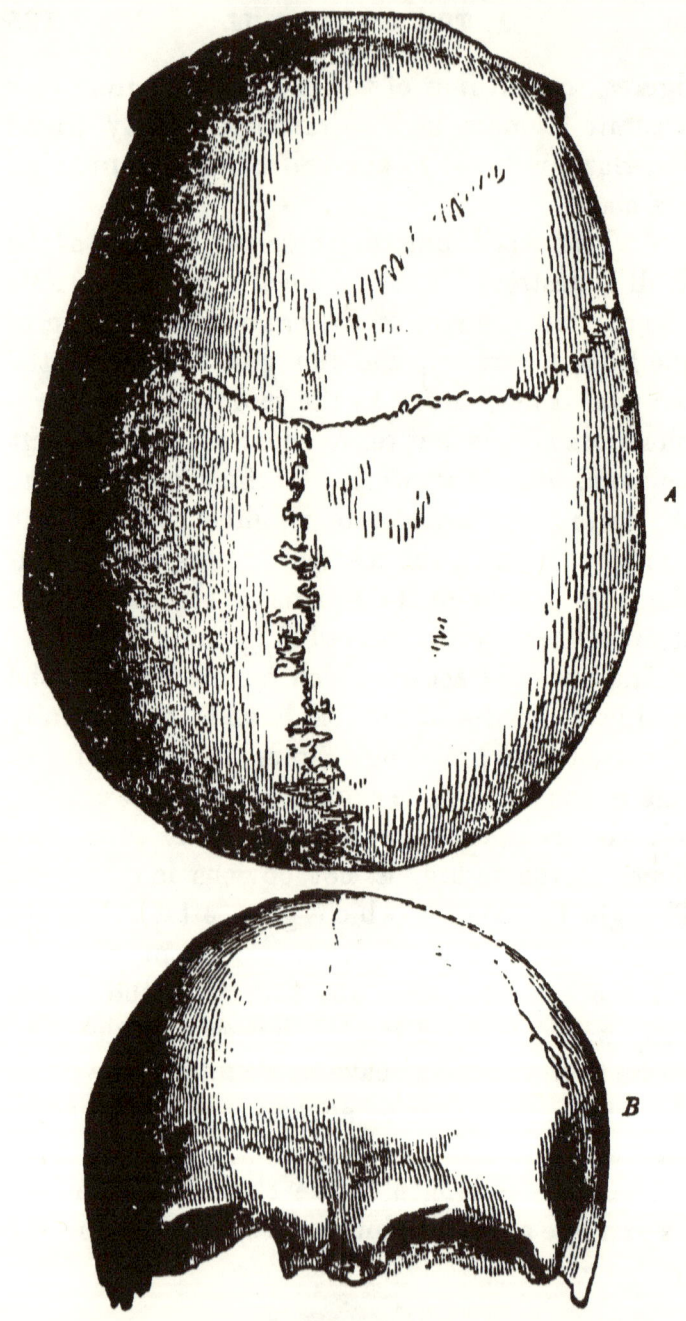

Fig. 24.—The Engis skull viewed from above (*A*) and in front (*B*).

with the interval between the parietal protuberances, is not more than 5·4 inches. The proportion of the length to the breadth is therefore very nearly as 100 to 70. If a line be drawn from the point at which the brow curves in towards the root of the nose, and which is called the "glabella" (*a*), (Fig. 23), to the occipital protuberance (*b*), and the distance to the highest point of the arch of the skull be measured perpendicularly from this line, it will be found to be 4·75 inches. Viewed from above, Fig. 24, *A*, the forehead presents an evenly rounded curve, and passes into the contour of the sides and back of the skull, which describes a tolerably regular elliptical curve.

The front view (Fig. 24, *B*) shows that the roof of the skull was very regularly and elegantly arched in the transverse direction, and that the transverse diameter was a little less below the parietal protuberances, than above them. The forehead cannot be called narrow in relation to the rest of the skull, nor can it be called a retreating forehead; on the contrary, the antero-posterior contour of the skull is well arched, so that the distance along that contour, from the nasal depression to the occipital protuberance, measures about 13·75 inches. The transverse arc of the skull, measured from one auditory foramen to the other, across the middle of the sagittal suture, is about 13 inches. The sagittal suture itself is 5·5 inches long.

The supraciliary prominences or brow-ridges

(on each side of *a*, Fig. 23) are well, but not excessively, developed, and are separated by a median depression. Their principal elevation is disposed so obliquely that I judge them to be due to large frontal sinuses.

If a line joining the glabella and the occipital protuberance (*a, b,* Fig. 23) be made horizontal, no part of the occipital region projects more than $\frac{1}{10}$th of an inch behind the posterior extremity of that line, and the upper edge of the auditory foramen (*c*) is almost in contact with a line drawn parallel with this upon the outer surface of the skull.

A transverse line drawn from one auditory foramen to the other traverses, as usual, the fore part of the occipital foramen. The capacity of the interior of this fragmentary skull has not been ascertained.

The history of the Human remains from the cavern in the Neanderthal may best be given in the words of their original describer, Dr. Schaaffhausen,[1] as translated by Mr. Busk.

"In the early part of the year 1857, a human skeleton was discovered in a limestone cave in the Neanderthal, near Hochdal, between Düsseldorf and Elberfeld. Of this, however, I was unable to procure more than a plaster cast of the cranium, taken at Elberfeld, from which I drew up an account of its

---

[1] *On the Crania of the most Ancient Races of Man.*—By Professor D. Schaaffhausen, of Bonn. (From Müller's *Archiv.*, 1858, pp. 453.) With Remarks, and original Figures, taken from a Cast of the Neanderthal Cranium. By George Busk, F.R.S., &c. *Natural History Review*, April, 1861.

remarkable conformation, which was, in the first instance, read on the 4th of February, 1857, at the meeting of the Lower Rhine Medical and Natural History Society, at Bonn.[1] Subsequently Dr. Fuhlrott, to whom science is indebted for the preservation of these bones, which were not at first regarded as human, and into whose possession they afterwards came, brought the cranium from Elberfeld to Bonn, and entrusted it to me for more accurate anatomical examination. At the General Meeting of the Natural History Society of Prussian Rhineland and Westphalia, at Bonn, on the 2nd of June, 1857,[2] Dr. Fuhlrott himself gave a full account of the locality, and of the circumstances under which the discovery was made. He was of opinion that the bones might be regarded as fossil; and in coming to this conclusion, he laid especial stress upon the existence of dendritic deposits, with which their surface was covered, and which were first noticed upon them by Professor Mayer. To this communication I appended a brief report on the results of my anatomical examination of the bones. The conclusions at which I arrived were: 1st. That the extraordinary form of the skull was due to a natural conformation hitherto not known to exist, even in the most barbarous races. 2nd. That these remarkable human remains belonged to a period antecedent to the time of the Celts and Germans, and were in all probability derived from one of the wild races of North-western Europe, spoken of by Latin writers; and which were encountered as autochthones by the German immigrants. And 3rdly. That it was beyond doubt that these human relics were traceable to a period at which the latest animals of the diluvium still existed; but that no proof of this assumption, nor consequently of their so-termed *fossil* condition, was afforded by the circumstances under which the bones were discovered.

"As Dr. Fuhlrott has not yet published his description of these circumstances, I borrow the following account of them from one of his letters. 'A small cave or grotto, high enough

---

[1] *Verhandl. d. Naturhist. Vereins der preuss. Rheinlande und Westphalens.*, xiv.—Bonn, 1857.
[2] *Ib.* Correspondenzblatt. No. 2.

to admit a man, and about 15 feet deep from the entrance, which is 7 or 8 feet wide, exists in the southern wall of the gorge of the Neanderthal, as it is termed, at a distance of about 100 feet from the Düssel, and about 60 feet above the bottom of the valley. In its earlier and uninjured condition, this cavern opened upon a narrow plateau lying in front of it, and from which the rocky wall descended almost perpendicularly into the river. It could be reached, though with difficulty, from above. The uneven floor was covered to a thickness of 4 or 5 feet with a deposit of mud, sparingly intermixed with rounded fragments of chert. In the removing of this deposit, the bones were discovered. The skull was first noticed, placed nearest to the entrance of the cavern; and further in, the other bones, lying in the same horizontal plane. Of this I was assured, in the most positive terms, by two labourers who were employed to clear out the grotto, and who were questioned by me on the spot. At first no idea was entertained of the bones being human; and it was not till several weeks after their discovery that they were recognised as such by me, and placed in security.

"'But, as the importance of the discovery was not at the time perceived, the labourers were very careless in the collecting, and secured chiefly only the larger bones; and to this circumstance it may be attributed that fragments merely of the probably perfect skeleton came into my possession.'

"My anatomical examination of these bones afforded the following results:—

"The cranium is of unusual size, and of a long-elliptical form. A most remarkable peculiarity is at once obvious in the extraordinary development of the frontal sinuses, owing to which the superciliary ridges, which coalesce completely in the middle, are rendered so prominent, that the frontal bone exhibits a considerable hollow or depression above, or rather behind them, whilst a deep depression is also formed in the situation of the root of the nose. The forehead is narrow and low, though the middle and hinder portions of the cranial arch are well developed. Unfortunately, the fragment of the skull that has been preserved consists only of the portion situated above the roof of the orbits and the superior occipital ridges, which are greatly developed, and almost conjoined so as to form a horizontal eminence. It includes almost the whole of the frontal bone, both parietals, a small part of the squamous and the upper-third of the occipital. The recently fractured surfaces show that the skull was broken at the time of its disinterment. The cavity holds 16,876 grains of water, whence its cubical contents may be estimated at 57·64 inches, or 1033·24 cubic centimetres. In making this estimation, the water is supposed to stand on a level with the orbital plate of the frontal, with the deepest notch in the squamous margin of the parietal, and with the superior semicircular ridges of the occipital. Estimated in dried millet-seed, the contents equalled 31 ounces, Prussian Apothecaries' weight. The semicircular line indicating the upper boundary of the attachment of the temporal muscle, though not very strongly marked, ascends nevertheless to more than half the height of the parietal bone. On the right superciliary ridge is observable an oblique furrow or depression, indicative of an injury received during life.[1] The coronal and sagittal sutures are on the exterior nearly closed, and on the inside so completely ossified as to have left no traces whatever, whilst the lambdoidal remains quite open. The depressions for the Pacchionian glands are deep and numerous; and there is an unusually deep vascular groove immediately behind the coronal suture, which, as it terminates in a foramen, no doubt transmitted a *vena emissaria*. The course of the frontal suture is indicated externally by a slight ridge; and where it joins the coronal, this ridge rises into a small protuberance. The course of the sagittal suture is grooved, and above the angle of the occipital bone the parietals are depressed.

	mm.[2]	inches.
The length of the skull from the nasal process of the frontal over the vertex to the superior semicircular lines of the occipital measures	303 (300)	= 12·0."

---

[1] This, Mr. Busk has pointed out, is probably the notch for the frontal nerve.

[2] The numbers in brackets are those which I should assign to the different measures, as taken from the plaster cast.—G. B.

	mm.	inches.
Circumference over the orbital ridges and the superior semicircular lines of the occipital	590 (590)	= 23·37" or 23".
Width of the frontal from the middle of the temporal line on one side to the same point on the opposite	104 (114)	= 4·1" — 4·5".
Length of the frontal from the nasal process to the coronal suture	133 (125)	= 5·25" — 5".
Extreme width of the frontal sinuses	25 (23)	= 1·0" — 0·9".
Vertical height above a line joining the deepest notches in the squamous border of the parietals	70	= 2·75".
Width of hinder part of skull from one parietal protuberance to the other	138 (150)	= 5·4" — 5·9".
Distance from the upper angle of the occipital to the superior semicircular lines	51 (60)	= 1·9" — 2·4".
Thickness of the bone at the parietal protuberance	8.	
————at the angle of the occipital	9.	
————at the superior semicircular line of the occipital	10	= 0·3".

"Besides the cranium, the following bones have been socured :—

"1. Both thigh-bones, perfect. These, like the skull, and all the other bones, are characterized by their unusual thickness, and the great development of all the elevations and depressions for the attachment of muscles. In the Anatomical Museum at Bonn, under the designation of 'Giant's-bones,' are some recent thigh-bones, with which in thickness the foregoing pretty nearly correspond, although they are shorter.

	Giant's bones.		Fossil bones.	
	mm.	inches.	mm.	inches.
Length	542	= 21·4"	438	= 17·4".
Diameter of head of femur	54	= 2·14"	53	= 2·0"

	Giant's bones.		Fossil bones.	
	mm.	inches.	mm.	inches
Diameter of lower articular end, from one condyle to the other	89	= 3·5″	87	= 3·4″
Diameter of femur in the middle	33	= 1·2″	30	= 1·1″

"2. A perfect right humerus, whose size shows that it belongs to the thigh-bones.

	mm.	inches.
Length	312	= 12·3′
Thickness in the middle	26	= 1·0″
Diameter of head	49	= 1·9″

" Also a perfect right radius of corresponding dimensions and the upper-third of a right ulna corresponding to the humerus and radius.

" 3. A left humerus, of which the upper-third is wanting, and which is so much slenderer than the right as apparently to belong to a distinct individual; a left *ulna*, which, though complete, is pathologically deformed, the coronoid process being so much enlarged by bony growth, that flexure of the elbow beyond a right angle must have been impossible; the anterior fossa of the humerus for the reception of the coronoid process being also filled up with a similar bony growth. At the same time, the olecranon is curved strongly downwards. As the bone presents no sign of rachitic degeneration, it may be supposed that an injury sustained during life was the cause of the anchylosis. When the left ulna is compared with the right radius, it might at first sight be concluded that the bones respectively belonged to different individuals, the ulna being more than half an inch too short for articulation with a corresponding radius. But it is clear that this shortening, as well as the attenuation of the left humerus, are both consequent upon the pathological condition above described.

"4. A left *ilium*, almost perfect, and belonging to the femur; a fragment of the right *scapula;* the anterior extremity of a rib of the right side; and the same part of a rib of the left side; the hinder part of a rib of the right side; and, lastly, two hinder portions and one middle portion of ribs which, from their unusually rounded shape, and abrupt curvature, more resemble the ribs of a carnivorous animal than those of a man. Dr. H. v. Meyer, however, to whose judgment I defer, will not venture to declare them to be ribs of any animal; and it only remains to suppose that this abnormal condition has arisen from an unusually powerful development of the thoracic muscles.

"The bones adhere strongly to the tongue, although, as proved by the use of hydrochloric acid, the greater part of the cartilage is still retained in them, which appears, however, to have undergone that transformation into gelatine which has been observed by v. Bibra in fossil bones. The surface of all the bones is in many spots covered with minute black specks, which, more especially under a lens, are seen to be formed of very delicate *dendrites*. These deposits, which were first observed on the bones by Dr. Mayer, are most distinct on the inner surface of the cranial bones. They consist of a ferruginous compound, and, from their black colour, may be supposed to contain manganese. Similar dendritic formations also occur, not unfrequently, on laminated rocks, and are usually found in minute fissures and cracks. At the meeting of the Lower Rhine Society at Bonn, on the 1st April, 1857, Prof. Mayer stated that he had noticed in the museum of Poppelsdorf similar dendritic crystallizations on several fossil bones of animals, and particularly on those of *Ursus spelœus*, but still more abundantly and beautifully displayed on the fossil bones and teeth of *Equus adamiticus*, *Elephas primigenius*, &c., from the caves of Bolve and Sundwig. Faint indications of similar *dendrites* were visible in a Roman skull from Siegburg; whilst other ancient skulls, which had lain for centuries in the earth, presented no trace of them.[1] I am indebted to H. v. Meyer for the following remarks on this subject:—

"'The incipient formation of dendritic deposits, which were formerly regarded as a sign of a truly fossil condition, is interesting. It has even been supposed that in diluvial deposits

---

[1] *Verh. des Naturhist. Vereins in Bonn*, xiv. 1857.

the presence of *dendrites* might be regarded as affording a certain mark of distinction between bones mixed with the diluvium at a somewhat later period and the true diluvial relics, to which alone it was supposed that these deposits were confined. But I have long been convinced that neither can the absence of *dendrites* be regarded as indicative of recent age, nor their presence as sufficient to establish the great antiquity of the objects upon which they occur. I have myself noticed upon paper, which could scarcely be more than a year old, dendritic deposits, which could not be distinguished from those on fossil bones. Thus I possess a dog's skull from the Roman colony of the neighbouring Heddersheim, *Castrum Hadrianum*, which is in no way distinguishable from the fossil bones from the Frankish caves; it presents the same colour, and adheres to the tongue just as they do; so that this character also, which, at a former meeting of German naturalists at Bonn, gave rise to amusing scenes between Buckland and Schmerling, is no longer of any value. In disputed cases, therefore, the condition of the bone can scarcely afford the means for determining with certainty whether it be fossil, that is to say, whether it belong to geological antiquity or to the historical period.'

"As we cannot now look upon the primitive world as representing a wholly different condition of things, from which no transition exists to the organic life of the present time, the designation of *fossil*, as applied to *a bone*, has no longer the sense it conveyed in the time of Cuvier. Sufficient grounds exist for the assumption that man coexisted with the animals found in the *diluvium*; and many a barbarous race may, before all historical time, have disappeared, together with the animals of the ancient world, whilst the races whose organization is improved have continued the genus. The bones which form the subject of this paper present characters which, although not decisive as regards a geological epoch, are, nevertheless, such as indicate a very high antiquity. It may also be remarked that, common as is the occurrence of diluvial animal bones in the muddy deposits of caverns, such remains have not hitherto been met with in the caves of the Neanderthal; and that the bones, which were covered by a deposit of mud not more than four or five feet thick, and without any protective covering of stalagmite, have retained the greatest part of their organic substance.

"These circumstances might be adduced against the probability of a geological antiquity. Nor should we be justified in regarding the cranial conformation as perhaps representing the most savage primitive type of the human race, since crania exist among living savages, which, though not exhibiting such a remarkable conformation of the forehead, which gives the skull somewhat the aspect of that of the large apes, still in other respects, as for instance in the greater depth of the temporal fossæ, the crest-like, prominent temporal ridges, and a generally less capacious cranial cavity, exhibit an equally low stage of development. There is no reason for supposing that the deep frontal hollow is due to any artificial flattening, such as is practised in various modes by barbarous nations in the Old and New World. The skull is quite symmetrical, and shows no indication of counter-pressure at the occiput, whilst, according to Morton, in the Flat-heads of the Columbia, the frontal and parietal bones are always unsymmetrical. Its conformation exhibits the sparing development of the anterior part of the head which has been so often observed in very ancient crania, and affords one of the most striking proofs of the influence of culture and civilization on the form of the human skull."

In a subsequent passage, Dr. Schaaffhausen remarks:

"There is no reason whatever for regarding the unusual development of the frontal sinuses in the remarkable skull from the Neanderthal as an individual or pathological deformity; it is unquestionably a typical race-character, and is physiologically connected with the uncommon thickness of the other bones of the skeleton, which exceeds by about one-half the usual proportions. This expansion of the frontal sinuses, which are appendages of the air-passages, also indicates an unusual force and power of endurance in the movements of the body, as may be concluded from the size of all the ridges and processes for the attachment of the muscles or bones. That this conclusion may be drawn from the existence of large frontal sinuses, and a prominence of the lower frontal region, is confirmed in many ways by other observations. By the same characters, according to Pallas, the wild horse is distinguished from the domesticated, and, according to Cuvier, the fossil cave-bear from every recent species of bear, whilst, according to Roulin, the pig, which has become wild in America, and regained a resemblance to the wild boar, is thus distinguished from the same animal in the domesticated state, as is the chamois from the goat; and, lastly, the bull-dog, which is characterised by its large bones and strongly-developed muscles from every other kind of dog. The estimation of the facial angle, the determination of which, according to Professor Owen, is also difficult in the great apes, owing to the very prominent supra-orbital ridges, in the present case is rendered still more difficult from the absence both of the auditory opening and of the nasal spine. But if the proper horizontal position of the skull be taken from the remaining portions of the orbital plates, and the ascending line made to touch the surface of the frontal bone behind the prominent supra-orbital ridges, the facial angle is not found to exceed 56°.[1] Unfortunately, no portions of the facial bones, whose conformation is so decisive as regards the form and expression of the head, have been preserved. The cranial capacity, compared with the uncommon strength of the corporeal frame, would seem to indicate a small cerebral development. The skull, as it is, holds about 31 ounces of millet-seed; and as, from the proportionate size of the wanting bones, the whole cranial cavity should have about 6 ounces more added, the contents, were it perfect, may be taken at 37 ounces. Tiedemann assigns, as the cranial contents in the Negro, 40, 38, and 35 ounces. The cranium holds rather more than 36 ounces of water which

---

[1] Estimating the facial angle in the way suggested, on the cast I should place it at 64° to 67°.—G. B.

corresponds to a capacity of 1033·24 cubic centimetres. Huschke estimates the cranial contents of a Negress at 1127 cubic centimetres; of an old Negro at 1146 cubic centimetres. The capacity of the Malay skulls, estimated by water, equalled 36, 33 ounces, whilst in the diminutive Hindoos it falls to as little as 27 ounces."

After comparing the Neanderthal cranium with many others, ancient and modern, Professor Schaaffhausen concludes thus:—

"But the human bones and cranium from the Neanderthal exceed all the rest in those peculiarities of conformation which lead to the conclusion of their belonging to a barbarous and savage race. Whether the cavern in which they were found, unaccompanied with any trace of human art, were the place of their interment, or whether, like the bones of extinct animals elsewhere, they had been washed into it, they may still be regarded as the most ancient memorial of the early inhabitants of Europe."

Mr. Busk, the translator of Dr. Schaaffhausen's paper, has enabled us to form a very vivid conception of the degraded character of the Neanderthal skull, by placing side by side with its outline, that of the skull of a Chimpanzee, drawn to the same absolute size.

Some time after the publication of the translation of Professor Schaaffhausen's Memoir, I was led to study the cast of the Neanderthal cranium with more attention than I had previously bestowed upon it, in consequence of wishing to supply Sir Charles Lyell with a diagram, exhibiting the special peculiarities of this skull, as compared with other human skulls. In order to do this it was necessary to identify, with precision, those points in the skulls compared which corresponded anatomically. Of these points, the glabella was obvious enough; but when I had distinguished another, defined by the occipital protuberance and superior semi-circular line, and had placed the outline of the Neanderthal skull against that of the Engis skull, in such a position that the glabella and occipital protuberance of both were intersected by the same straight line, the difference was so vast and the flattening of the Neanderthal skull so prodigious (compare Figs. 23 and 25 A), that I at first imagined I must have fallen into some error. And I was the more inclined to suspect this, as, in ordinary human skulls, the occipital protuberance and superior semicircular curved line on the exterior of the occiput correspond pretty closely with the "lateral sinuses" and the line of attachment of the tentorium internally. But on the tentorium rests, as I have said in the preceding Essay, the posterior lobe of the brain; and hence, the occipital protuberance, and the curved line in question, indicate, approximately, the lower limits of that lobe. Was it possible for a human being to have the brain thus flattened and depressed; or, on the other hand, had the muscular ridges shifted their position? In order to solve these doubts, and to decide the question whether the great supraciliary projections did, or

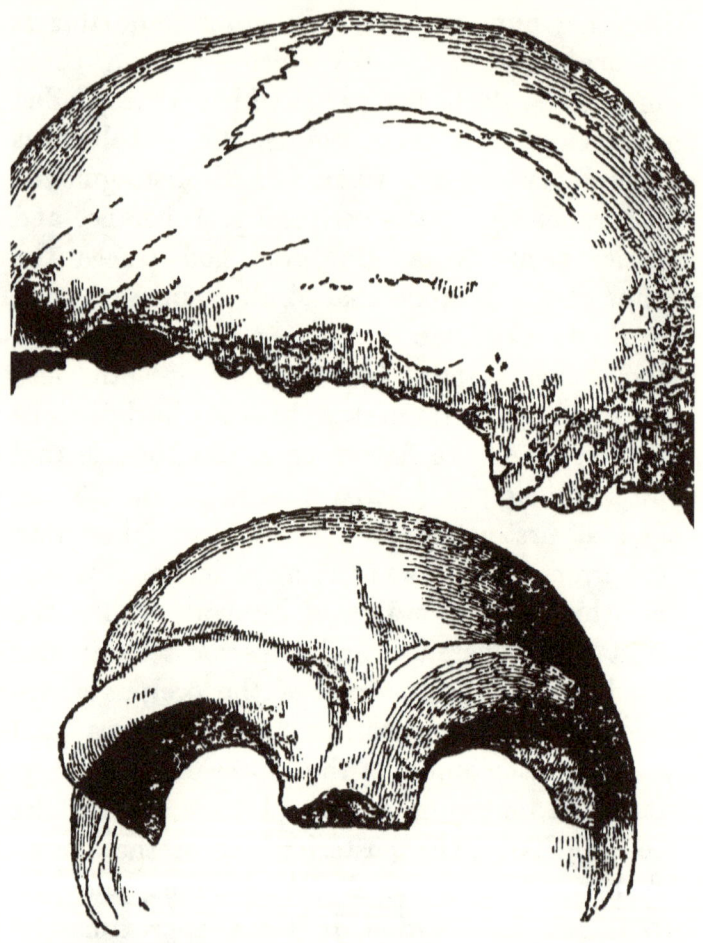

Fig. 25.—The skull from the Neanderthal cavern. A, side, outlines from camera lucida drawings, one half the natural size, photographs. *a* glabella ; *b* occipital protuberance ; *d* lamb-

did not, arise from the development of the frontal sinuses, I requested Sir Charles Lyell to be so good as to obtain for me from Dr. Fuhlrott,

THE NEANDERTHAL MAN 181

the possessor of the skull, answers to certain queries, and if possible a cast, or at any rate drawings, or photographs, of the interior of the skull.

Dr. Fuhlrott replied, with a courtesy and

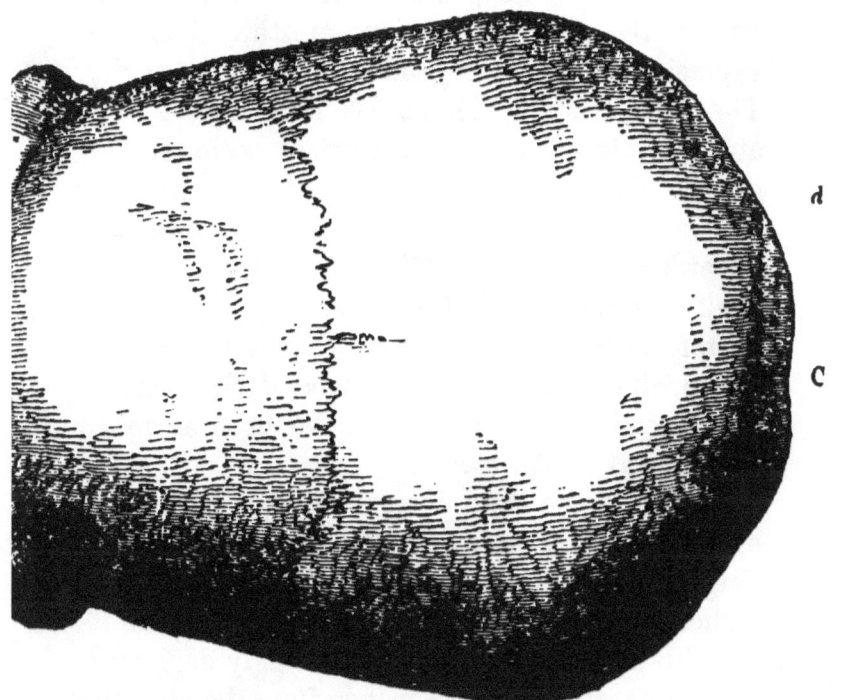

B, front, and C, top view. One half the natural size. The by Mr. Busk: the details from the cast and from Dr. Fuhlrott's doidal suture.

readiness for which I am infinitely indebted to him, to my inquiries, and furthermore sent three excellent photographs. One of these gives a side view of the skull, and from it Fig. 25 A has been shaded. The second (Fig. 26 A) exhibits the wide openings of the frontal sinuses upon the inferior surface of the frontal part of the skull, into which, Dr. Fuhlrott writes, "a probe may be introduced to the depth of an inch," and demonstrates the great extension of the thickened supraciliary ridges beyond the cerebral cavity. The third, lastly (Fig. 26 B), exhibits the edge and the interior of the posterior, or occipital, part of the skull, and shows very clearly the two depressions for the lateral sinuses, sweeping inwards towards the middle line of the roof of the skull, to form the longitudinal sinus. It was clear, therefore, that I had not erred in my interpretation, and that the posterior lobe of the brain of the Neanderthal man must have been as much flattened as I suspected it to be.

In truth, the Neanderthal cranium has most extraordinary characters. It has an extreme length of 8 inches, while its breadth is only 5·75 inches, or, in other words, its length is to its breadth as 100 : 72. It is exceedingly depressed, measuring only about 3·4 inches from the glabello-occipital line to the vertex. The longitudinal arc, measured in the same way as in the Engis skull, is 12 inches; the transverse arc cannot be exactly ascertained, in consequence of the absence of the temporal bones, but was probably about the same, and certainly exceeded 10¼ inches. The hori zontal circumference is 23 inches. But this great circumference arises largely from the vast de-

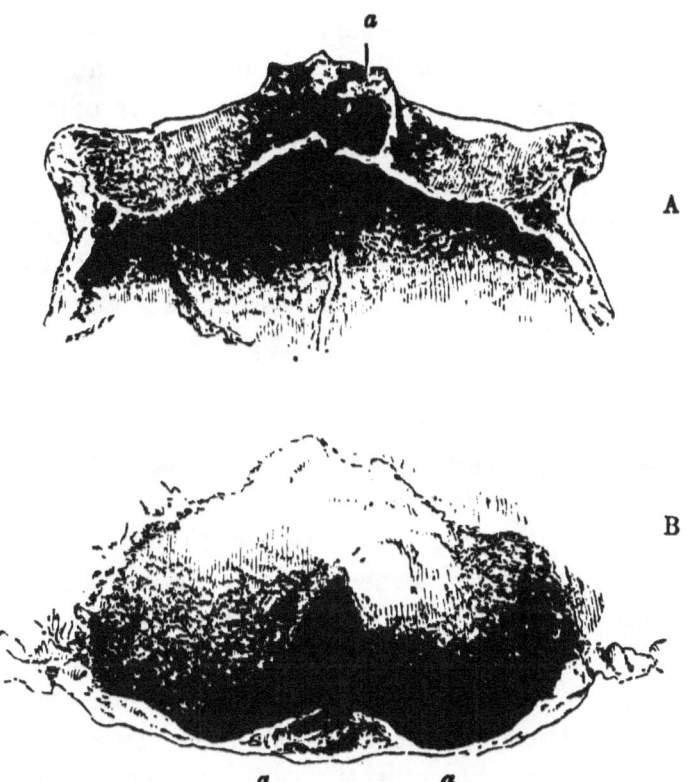

Fig. 26.—Drawings from Dr. Fuhlrott's photographs of parts of the interior of the Neanderthal cranium. A view of the under and inner surface of the frontal region, showing the inferior apertures of the frontal sinuses (*a*). B corresponding view of the occipital region of the skull, showing the impressions of the lateral sinuses (*aa*).

velopment of the supraciliary ridges, though the perimeter of the brain case itself is not small.

The large supraciliary ridges give the forehead a far more retreating appearance than its internal contour would bear out.

To an anatomical eye, the posterior part of the skull is even more striking than the anterior. The occipital protuberance occupies the extreme posterior end of the skull, when the glabello-occipital line is made horizontal, and so far from any part of the occipital region extending beyond it, this region of the skull slopes obliquely upward and forward, so that the lambdoidal suture is situated well upon the upper surface of the cranium. At the same time, notwithstanding the great length of the skull, the sagittal suture is remarkably short (4½ inches), and the squamosal suture is very straight.

In reply to my questions Dr. Fuhlrott writes that the occipital bone " is in a state of perfect preservation as far as the upper semicircular line, which is a very strong ridge, linear at its extremities, but enlarging towards the middle, where it forms two ridges (bourrelets), united by a linear continuation, which is slightly depressed in the middle."

"Below the left ridge the bone exhibits an obliquely inclined surface, six lines (French) long, and twelve lines wide."

This last must be the surface, the contour of which is shown in Fig. 25 A, below *b*. It is particularly interesting, as it suggests that, notwithstanding the flattened condition of the occiput, the posterior cerebral lobes must have projected considerably beyond the cerebellum, and as it constitutes one among several points of similarity between the Neanderthal cranium and certain Australian skulls.

Such are the two best known forms of human cranium, which have been found in what may be fairly termed a fossil state. Can either be shown to fill up or diminish, to any appreciable extent, the structural interval which exists between Man and the man-like apes? Or, on the other hand, does neither depart more widely from the average structure of the human cranium, than normally formed skulls of men are known to do at the present day?

It is impossible to form any opinion on these questions, without some preliminary acquaintance with the range of variation exhibited by human structure in general—a subject which has been but imperfectly studied, while even of what is known, my limits will necessarily allow me to give only a very imperfect sketch.

The student of anatomy is perfectly well aware that there is not a single organ of the human body the structure of which does not vary, to a greater or less extent, in different individuals. The skeleton varies in the proportions, and even to a certain extent in the connexions, of its constituent bones. The muscles which move the bones vary largely in their attachments. The varieties in the mode of distribution of the arteries are carefully classified, on account of the practical importance of a knowledge of their shiftings to the surgeon. The characters of the brain vary immensely, nothing being less constant than the form and size of the cerebral hemispheres, and the richness of the convolutions upon their surface, while the most changeable structures of all in the human brain are exactly those on which the unwise attempt has been made to base the distinctive characters of humanity, viz. the posterior cornu of the lateral ventricle, the hippocampus minor, and the degree of projection of the posterior lobe beyond the cerebellum. Finally, as all the world knows, the hair and skin of human beings may present the most extraordinary diversities in colour and in texture.

So far as our present knowledge goes, the majority of the structural varieties to which allusion is here made, are individual. The ape-like arrangement of certain muscles which is occasionally met with[1] in the white races of mankind, is not known to be more common among Negroes or Australians: nor because the brain of the Hottentot Venus was found to be smoother, to have its convolutions more symmetrically disposed,

[1] See an excellent Essay by Mr. Church on the Myology of the Orang, in the *Natural History Review* for 1861.

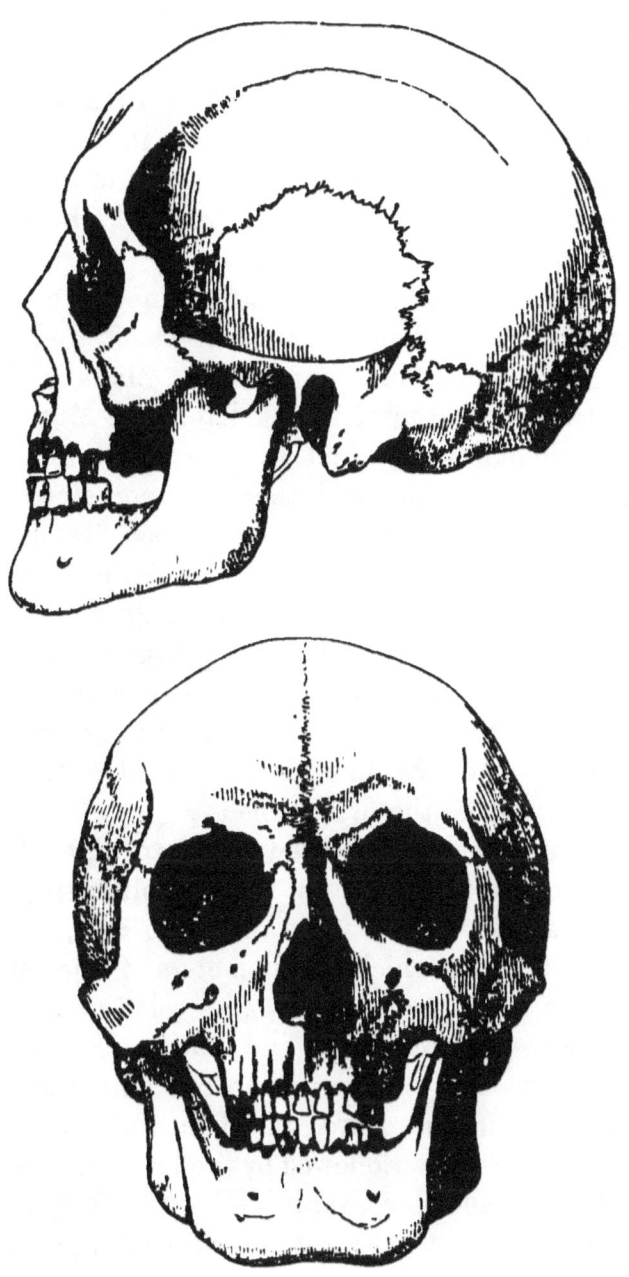

Fig. 27.—Side and front views of the round and orthognathous skull of a Calmuck after Von Baer. One-third the natural size.

and to be, so far, more ape-like than that of ordinary Europeans, are we justified in concluding a like condition of the brain to prevail universally among the lower races of mankind, however probable that conclusion may be.

We are, in fact, sadly wanting in information respecting the disposition of the soft and destructible organs of every Race of Mankind but our own; and even of the skeleton, our Museums are lamentably deficient in every part but the cranium. Skulls enough there are, and since the time when Blumenbach and Camper first called attention to the marked and singular differences which they exhibit, skull collecting and skull measuring has been a zealously pursued branch of Natural History, and the results obtained have been arranged and classified by various writers, among whom the late active and able Retzius must always be the first named.

Human skulls have been found to differ from one another, not merely in their absolute size and in the absolute capacity of the brain case, but in the proportions which the diameters of the latter bear to one another; in the relative size of the bones of the face (and more particularly of the jaws and teeth) as compared with those of the skull; in the degree to which the upper jaw (which is of course followed by the lower) is thrown backwards and downwards under the forepart of the brain case, or forwards and upwards in front of and beyond it. They differ further in the relations of the transverse diameter of the face, taken through the cheek bones, to the transverse diameter of the skull; in the more rounded or more gable-like form of the roof of the skull, and in the degree to which the hinder part of the skull is flattened or projects beyond the ridge, into and below which the muscles of the neck are inserted.

In some skulls the brain case may be said to be "*round*," the extreme length not exceeding the extreme breadth by a greater proportion than 100 to 80, while the difference may be much less.[1] Men possessing such skulls were termed by Retzius "*brachycephalic*," and the skull of a Calmuck, of which a front and side view (reduced outline copies of which are given in Figure 27) are depicted by Von Baer in his excellent "Crania selecta," affords a very admirable sample of that kind of skull. Other skulls, such as that of a Negro copied in Fig. 28 from Mr. Busk's "Crania typica," have a very different, greatly elongated form, and may be termed "*oblong*." In this skull the extreme length is to the extreme breadth as 100 to not more than 67, and the transverse diameter of the human skull may fall below even this proportion. People having such skulls were called by Retzius "*dolichocephalic*."

The most cursory glance at the side views of

[1] In no normal human skull does the breadth of the brain-case exceed its length.

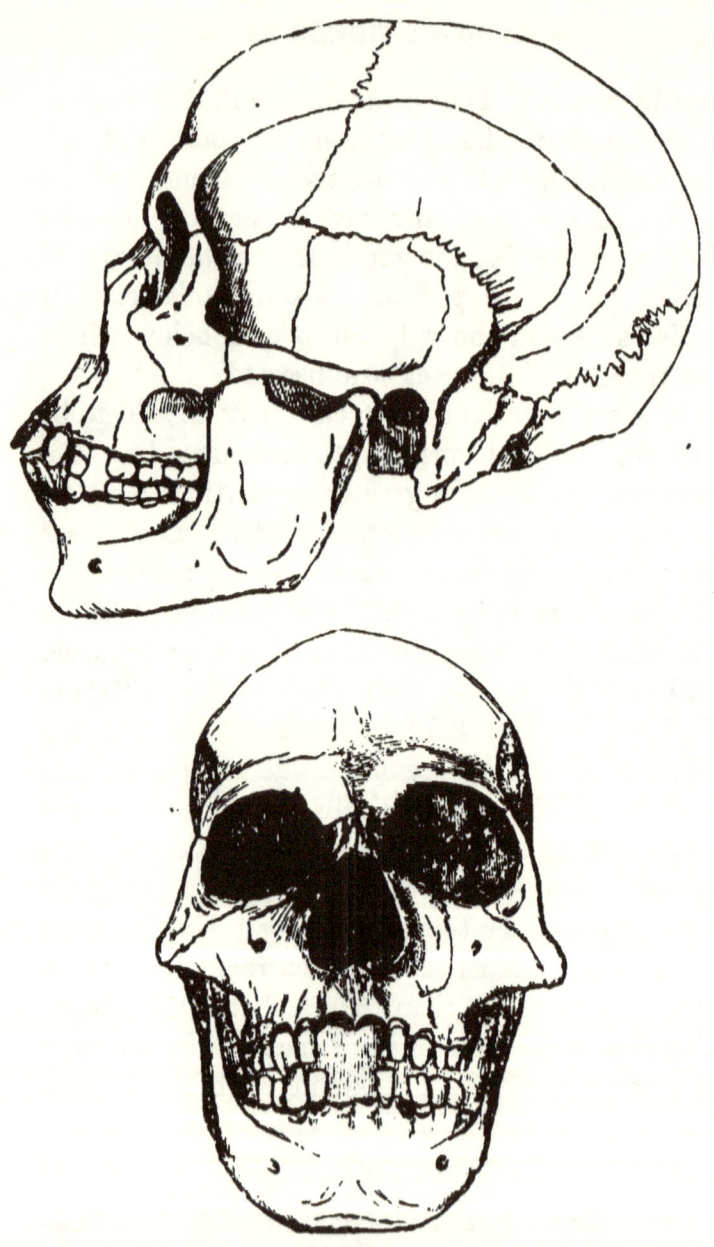

FIG. 28.—Oblong and prognathous skull of a Negro; side and front views. One-third of the natural size.

these two skulls will suffice to prove that they differ, in another respect, to a very striking extent. The profile of the face of the Calmuck is almost vertical, the facial bones being thrown downwards and under the fore part of the skull. The profile of the face of the Negro, on the other hand, is singularly inclined, the front part of the jaws projecting far forward beyond the level of the fore part of the skull. In the former case the skull is said to be "*orthognathous*" or straight-jawed; in the latter, it is called "*prognathous*," a term which has been rendered, with more force than elegance, by the Saxon equivalent,—"snouty."

Various methods have been devised in order to express with some accuracy the degree of prognathism or orthognathism of any given skull; most of these methods being essentially modifications of that devised by Peter Camper, in order to attain what he called the "facial angle."

But a little consideration will show that any "facial angle" that has been devised, can be competent to express the structural modifications involved in prognathism and orthognathism, only in a rough and general sort of way. For the lines, the intersection of which forms the facial angle, are drawn through points of the skull, the position of each of which is modified by a number of circumstances, so that the angle obtained is a complex resultant of all these circumstances, and is not the expression of any one definite organic relation of the parts of the skull.

I have arrived at the conviction that no comparison of crania is worth very much that is not founded upon the establishment of a relatively fixed base line, to which the measurements, in all cases, must be referred. Nor do I think it is a very difficult matter to decide what that base line should be. The parts of the skull, like those of the rest of the animal framework, are developed in succession: the base of the skull is formed before its sides and roof; it is converted into cartilage earlier and more completely than the sides and roof: and the cartilaginous base ossifies, and becomes soldered into one piece long before the roof. I conceive then that the base of the skull may be demonstrated developmentally to be its relatively fixed part, the roof and sides being relatively movable.

The same truth is exemplified by the study of the modifications which the skull undergoes in ascending from the lower animals up to man.

In such a mammal as a Beaver (Fig. 29), a line ($a\ b$) drawn through the bones, termed basioccipital, basisphenoid, and presphenoid, is very long in proportion to the extreme length of the cavity which contains the cerebral hemispheres ($g\ h$). The plane of the occipital foramen ($b\ c$) forms a slightly acute angle with this "basicranial axis," while the plane of the tentorium ($i\ T$) is inclined at rather more than 90° to the "basicranial axis"; and so is the plane of the perforated plate ($a\ d$), by which the filaments of the olfactory nerve

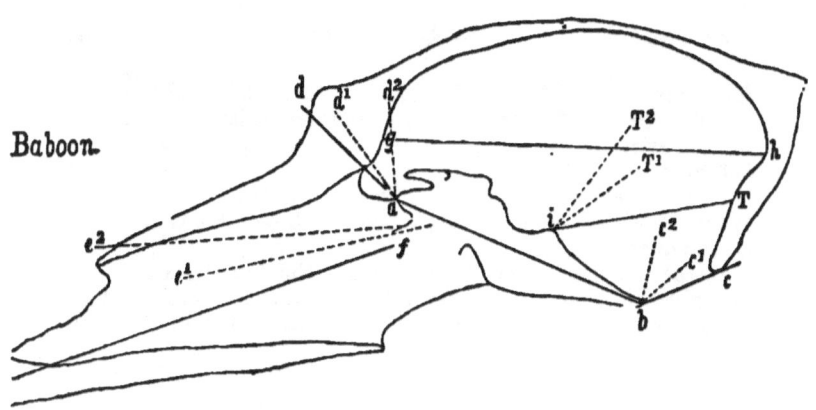

Fig. 29.—Longitudinal and vertical sections of the skulls of a Beaver (*Castor Canadensis*), a Lemur (*L. Catta*), and a Baboon (*Cynocephalus Papio*), ab, the basicranial axis; bc, the occipital plane; iT, the tentorial plane; ad, the olfactory plane; fe, the basifacial axis; cba, occipital angle; Tia, tentorial angle; dab, olfactory angle; efb, cranio-facial angle; gh, extreme length of the cavity which lodges the cerebral hemispheres or "cerebral length." The length of the basicranial axis as to this length, or, in other words, the proportional length of the line gh to that of ab taken as 100, in the three skulls, is as

follows:—Beaver, 70 to 100; Lemur, 119 to 100; Baboon, 144 to 100. In an adult male Gorilla the cerebral length is as 170 to the basicranial axis taken as 100, in the Negro (Fig. 30) as 236 to 100. In the Constantinople skull (Fig. 30) it is as 266 to 100. The difference between the highest Ape's skull and the lowest Man's is therefore very strikingly brought out by these measurements.

In the diagram of the Baboon's skull the dotted lines d^1 d^2, &c., give the angles of the Lemur's and Beaver's skull, as laid down upon the basicranial axis of the Baboon. The line $a\,b$ has the same length in each diagram.

leave the skull. Again, a line drawn through the axis of the face, between the bones called ethmoid and vomer—the "basifacial axis" ($f.\,e.$) forms an exceedingly obtuse angle, where, when produced, it cuts the "basicranial axis."

If the angle made by the line $b\,c$ with $a\,b$, be called the "occipital angle," and the angle made by the line $a\,d$ with $a\,b$ be termed the "olfactory angle" and that made by $i\,T$ with $a\,b$ the "tentorial angle" then all these, in the mammal in question, are nearly right angles, varying between 80° and 110°. The angle $e\,f\,b$, or that made by the cranial with the facial axis, and which may be termed the "cranio-facial angle," is extremely obtuse, amounting, in the case of the Beaver, to at least 150°.

But if a series of sections of mammalian skulls, intermediate between a Rodent and a Man (Fig. 29), be examined, it will be found that in the higher crania the basi-cranial axis becomes shorter relatively to the cerebral length; that the "olfactory angle" and "occipital angle" become more obtuse; and that the "cranio-facial angle," becomes more acute by the bending down, as it were, of the facial axis upon the cranial axis. At the same time, the roof of the cranium becomes more and more arched, to allow of the increasing height of the cerebral hemispheres, which is eminently characteristic of man, as well as of that backward extension, beyond the cerebellum, which reaches its maximum in the South American Monkeys. So that, at last, in the human skull (Fig. 30), the cerebral length is between twice and thrice as great as the length of the basicranial axis; the olfactory plane is 20° or 30° on the *under* side of that axis; the occipital angle, instead of being less than 90°, is as much as 150° or 160°; the cranio-facial angle may be 90° or less, and the vertical height of the skull may have a large proportion to its length.

It will be obvious, from an inspection of the diagrams, that the basicranial axis is, in the ascending series of Mammalia, a relatively fixed line, on which the bones of the sides and roof of the cranial cavity, and of the face, may be said to revolve downwards and forwards or backwards, according to their position. The arc described by any one bone or plane, however, is not by any means always in proportion to the arc described by another.

Now comes the important question, can we

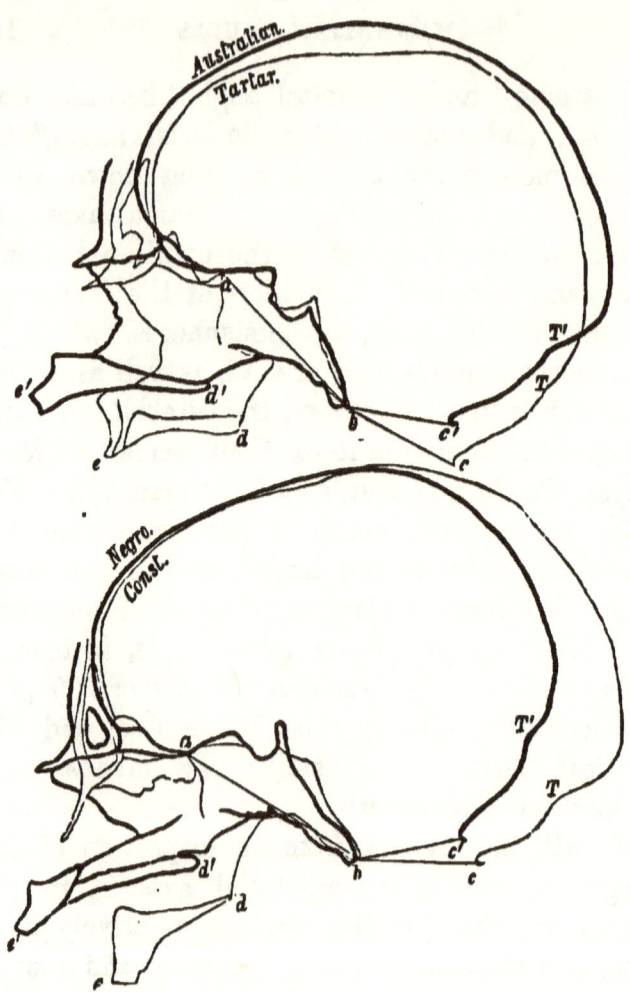

Fig. 30.—Sections of orthognathous (light contour) and prognathous (dark contour) skulls, one-third of the natural size. *a b*, Basicranial axis; *b c*, *b' c'*, plane of the occipital foramen; *d d'*, hinder end of the palatine bone; *e e'*, front end of the upper jaw; *T T'*, insertion of the tentorium.

discern, between the lowest and the highest forms of the human cranium anything answering, in however slight a degree, to this revolution of the side and roof bones of the skull upon the basi-cranial axis observed upon so great a scale in the mammalian series? Numerous observations lead me to believe that we must answer this question in the affirmative.

The diagrams in Figure 30 are reduced from very carefully made diagrams of sections of four skulls, two round and orthognathous, two long and prognathous, taken longitudinally and vertically, through the middle. The sectional diagrams have then been superimposed, in such a manner, that the basal axes of the skulls coincide by their anterior ends, and in their direction. The deviations of the rest of the contours (which represent the interior of the skulls only) show the differences of the skulls from one another, when these axes are regarded as relatively fixed lines.

The dark contours are those of an Australian and of a Negro skull: the light contours are those of a Tartar skull, in the Museum of the Royal College of Surgeons; and of a well developed round skull from a cemetery in Constantinople, of uncertain race, in my own possession.

It appears, at once, from these views, that the prognathous skulls, so far as their jaws are concerned, do really differ from the orthognathous in much the same way as, though to a far less degree than, the skulls of the lower mammals differ from those of Man. Furthermore, the plane of the occipital foramen (*b c*) forms a somewhat smaller angle with the axis in these particular prognathous skulls than in the orthognathous; and the like may be slightly true of the perforated plate of the ethmoid—though this point is not so clear. But it is singular to remark that, in another respect, the prognathous skulls are less ape-like than the orthognathous, the cerebral cavity projecting decidedly more beyond the anterior end of the axis in the prognathous, than in the orthognathous, skulls.

It will be observed that these diagrams reveal an immense range of variation in the capacity and relative proportion to the cranial axis, of the different regions of the cavity which contains the brain, in the different skulls. Nor is the difference in the extent to which the cerebral overlaps the cerebellar cavity less singular. A round skull (Fig. 30, *Const.*) may have a greater posterior cerebral projection than a long one (Fig. 30, *Negro*).

Until human crania have been largely worked out in a manner similar to that here suggested—until it shall be an opprobrium to an ethnological collection to possess a single skull which is not bisected longitudinally—until the angles and measurements here mentioned, together with a number of others of which I cannot speak in this place, are determined, and tabulated with reference to the basicranial axis as unity, for large numbers of skulls of the different races of Mankind, I do not think we shall have any very safe basis for that ethnological craniology which aspires to give the anatomical characters of the crania of the different Races of Mankind.

At present, I believe that the general outlines of what may be safely said upon that subject may be summed up in a very few words. Draw a line on a globe, from the Gold Coast in Western Africa to the steppes of Tartary. At the southern and western end of that line there live the most dolichocephalic, prognathous, curly-haired, dark-skinned of men—the true Negroes. At the northern and eastern end of the same line there live the most brachycephalic, orthognathous, straight-haired, yellow-skinned of men—the Tartars and Calmucks. The two ends of this imaginary line are indeed, so to speak, ethnological antipodes. A line drawn at right angles, or nearly so, to this polar line through Europe and Southern Asia to Hindostan, would give us a sort of equator, around which round-headed, oval-headed, and oblong-headed, prognathous and orthognathous, fair and dark races—but none possessing the excessively marked characters of Calmuck or Negro—group themselves.

It is worthy of notice that the regions of the

antipodal races are antipodal in climate, the greatest contrast the world affords, perhaps, being that between the damp, hot, steaming, alluvial coast plains of the West Coast of Africa and the arid, elevated steppes and plateaux of Central Asia, bitterly cold in winter, and as far from the sea as any part of the world can be.

From Central Asia eastward to the Pacific Islands and subcontinents on the one hand, and to America on the other, brachycephaly and orthognathism gradually diminish, and are replaced by dolichocephaly and prognathism, less, however, on the American Continent (throughout the whole length of which a rounded type of skull prevails largely, but not exclusively)[1] than in the Pacific region, where, at length, on the Australian Continent and in the adjacent islands, the oblong skull, the projecting jaws, and the dark skin reappear; with so much departure, in other respects, from the Negro type, that ethnologists assign to these people the special title of "Negritoes."

The Australian skull is remarkable for its narrowness and for the thickness of its walls, especially in the region of the supraciliary ridge, which is frequently, though not by any means invariably, solid throughout, the frontal sinuses remaining undeveloped. The nasal depression,

---

[1] See Dr. D. Wilson's valuable paper "On the supposed prevalence of one Cranial Type throughout the American Aborigines."—*Canadian Journal*, Vol. II. 1857.

again, is extremely sudden, so that the brows overhang and give the countenance a particularly lowering, threatening expression. The occipital region of the skull, also, not unfrequently becomes less prominent; so that it not only fails to project beyond a line drawn perpendicular to the hinder extremity of the glabello-occipital line, but even, in some cases, begins to shelve away from it, forwards, almost immediately. In consequence of this circumstance, the parts of the occipital bone which lie above and below the tuberosity make a much more acute angle with one another than is usual, whereby the hinder part of the base of the skull appears obliquely truncated. Many Australian skulls have a considerable height, quite equal to that of the average of any other race, but there are others in which the cranial roof becomes remarkably depressed, the skull, at the same time, elongating so much that, probably, its capacity is not diminished. The majority of skulls possessing these characters, which I have seen, are from the neighbourhood of Port Adelaide in South Australia, and have been used by the natives as water vessels; to which end the face has been knocked away, and a string passed through the vacuity and the occipital foramen, so that the skull was suspended by the greater part of its basis.

Figure 31 represents the contour of a skull of this kind from Western Port, with the jaw attached, and of the Neanderthal skull, both reduced to one-third of the size of nature. A small additional amount of flattening and lengthening, with a corresponding increase of the supraciliary

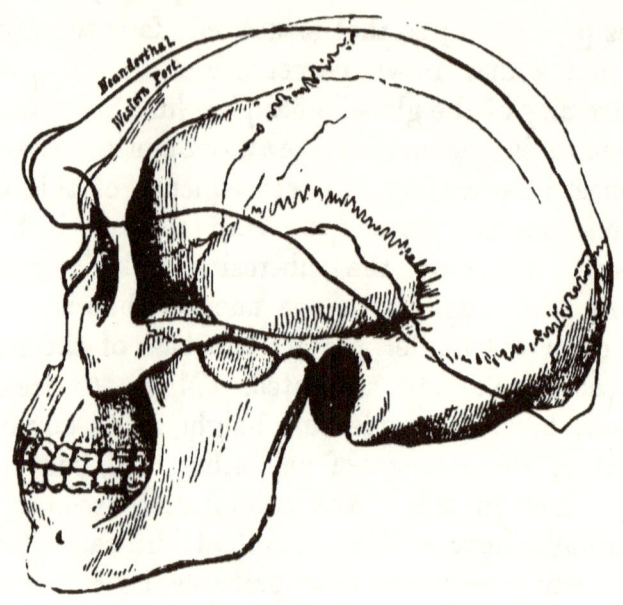

Fig. 31.—An Australian skull from Western Port, in the Museum of the Royal College of Surgeons, with the contour of the Neanderthal skull. Both reduced to one-third the natural size.

ridge, would convert the Australian brain case into a form identical with that of the aberrant fossil.

And now, to return to the fossil skulls, and to the rank which they occupy among, or beyond, these existing varieties of cranial conformation. In the first place, I must remark, that, as Professor Schmerling well observed (*supra, p.* 161) in commenting upon the Engis skull, the formation of a safe judgment upon the question is greatly hindered by the absence of the jaws from both the crania, so that there is no means of deciding, with certainty, whether they were more or less prognathous than the lower existing races of mankind. And yet, as we have seen, it is more in this respect than any other, that human skulls vary, towards and from, the brutal type—the brain case of an average dolichocephalic European differing far less from that of a Negro, for example, than his jaws do. In the absence of the jaws, then, any judgment on the relations of the fossil skulls to recent Races must be accepted with a certain reservation.

But taking the evidence as it stands, and turning first to the Engis skull, I confess I can find no character in the remains of that cranium which, if it were a recent skull, would give any trustworthy clue as to the Race to which it might appertain. Its contours and measurements agree very well with those of some Australian skulls which I have examined—and especially has it a tendency towards that occipital flattening, to the great extent of which, in some Australian skulls, I have alluded. But all Australian skulls do not present this flattening, and the supraciliary ridge of the Engis skull is quite unlike that of the typical Australians.

On the other hand, its measurements agree equally well with those of some European skulls. And assuredly, there is no mark of degradation about any part of its structure. It is, in fact, a fair average human skull, which might have belonged to a philosopher, or might have contained the thoughtless brains of a savage.

The case of the Neanderthal skull is very different. Under whatever aspect we view this cranium, whether we regard its vertical depression, the enormous thickness of its supraciliary ridges, its sloping occiput, or its long and straight squamosal suture, we meet with ape-like characters, stamping it as the most pithecoid of human crania yet discovered. But Professor Schaaffhausen states (*supra*, p. 178), that the cranium, in its present condition, holds 1033·24 cubic centimetres of water, or about 63 cubic inches, and as the entire skull could hardly have held less than an additional 12 cubic inches, its capacity may be estimated at about 75 cubic inches, which is the average capacity given by Morton for Polynesian and Hottentot skulls.

So large a mass of brain as this, would alone suggest that the pithecoid tendencies, indicated by this skull, did not extend deep into the organization; and this conclusion is borne out by the dimensions of the other bones of the skeleton given by Professor Schaaffhausen, which show that the absolute height and relative proportions of the limbs, were quite those of an European of middle stature. The bones are indeed stouter, but this and the great development of the muscular ridges noted by Dr. Schaaffhausen, are characters to be expected in savages. The Patagonians, exposed without shelter or protection to a climate possibly not very dissimilar from that of Europe at the time during which the Neanderthal man lived, are remarkable for the stoutness of their limb bones.

In no sense, then, can the Neanderthal bones be regarded as the remains of a human being intermediate between Men and Apes. At most, they demonstrate the existence of a Man whose skull may be said to revert somewhat towards the pithecoid type—just as a Carrier, or a Pouter, or a Tumbler, may sometimes put on the plumage of its primitive stock, the *Columba livia*. And indeed, though truly the most pithecoid of known human skulls, the Neanderthal cranium is by no means so isolated as it appears to be at first, but forms, in reality, the extreme term of a series leading gradually from it to the highest and best developed of human crania. On the one hand, it is closely approached by the flattened Australian skulls, of which I have spoken, from which other Australian forms lead us gradually up to skulls having very much the type of the Engis cranium.

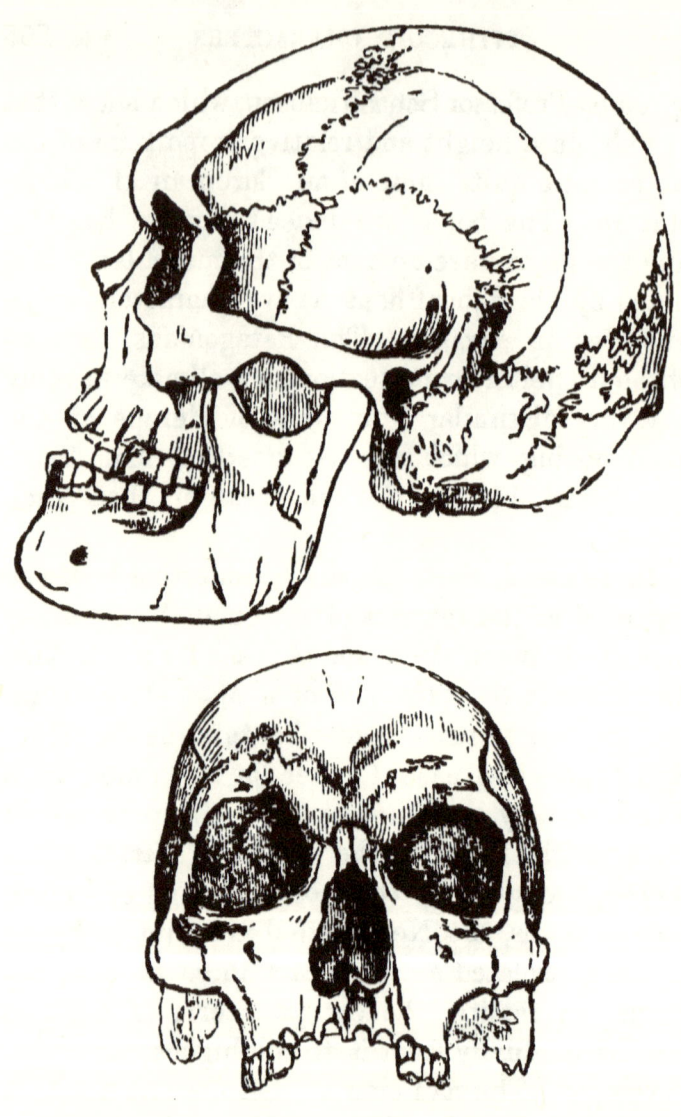

FIG. 32.—Ancient Danish skull from a tumulus at Borreby; one-third of the natural size. From a camera lucida drawing by Mr. Busk.

And, on the other hand, it is even more closely affined to the skulls of certain ancient people who inhabited Denmark during the "stone period," and were probably either contemporaneous with, or later than, the makers of the "refuse heaps," or "Kjokkenmöddings" of that country.

The correspondence between the longitudinal contour of the Neanderthal skull and that of some of those skulls from the tumuli at Borreby, very accurate drawings of which have been made by Mr. Busk, is very close. The occiput is quite as retreating, the supraciliary ridges are nearly as prominent, and the skull is as low. Furthermore, the Borreby skull resembles the Neanderthal form more closely than any of the Australian skulls do, by the much more rapid retrocession of the forehead. On the other hand, the Borreby skulls are all somewhat broader, in proportion to their length, than the Neanderthal skull, while some attain that proportion of breadth to length (80 : 100) which constitutes brachycephaly.[1]

In conclusion, I may say, that the fossil remains of Man hitherto discovered do not seem to me to

[1 For a further discussion of the characters of the Neanderthal skull, see "Natural History Review," 1864. I there say (p. 443): "That the Neanderthal skull exhibits the lowest type of human cranium at present known, so far as it presents certain pithecoid characters in a more exaggerated form than any other: but that, inasmuch as a complete series of gradations can be found, among recent human skulls, between it and the best developed forms, there is no ground for separating its pos-

take us appreciably nearer to that lower pithecoid form, by the modification of which he has, probably, become what he is. And considering what is now known of the most ancient Races of men; seeing that they fashioned flint axes and flint knives and bone-skewers, of much the same pattern as those fabricated by the lowest savages at the present day, and that we have every reason to believe the habits and modes of living of such people to have remained the same from the time of the Mammoth and the tichorhine Rhinoceros till now, I do not know that this result is other than might be expected.

Where, then, must we look for primæval Man? Was the oldest *Homo sapiens* pliocene or miocene, or yet more ancient? In still older strata do the fossilized bones of an ape more anthropoid, or a Man more pithecoid, than any yet known await the researches of some unborn paleontologist?

Time will show. But, in the meanwhile, if any form of the doctrine of progressive development is correct, we must extend by long epochs the most liberal estimate that has yet been made of the antiquity of Man.

sessor specifically, still less generically, from *Homo sapiens*. At present, we have no sufficient warranty for declaring it to be either the type of a distinct race, or a member of any existing one; nor do the anatomical characters of the skull justify any conclusion as to the age to which it belongs." See also the essay on the Aryan question in this volume. 1894.]

IV

ON THE METHODS AND RESULTS OF ETHNOLOGY

[1865]

ETHNOLOGY is the science which determines the distinctive characters of the persistent modifications of mankind; which ascertains the distribution of those modifications in present and past times, and seeks to discover the causes, or conditions of existence, both of the modifications and of their distribution. I say "persistent" modifications, because, unless incidentally, ethnology has nothing to do with chance and transitory peculiarities of human structure. And I speak of "persistent modifications" or "stocks" rather than of "varieties," or "races," or "species," because each of these last well-known terms implies, on the part of its employer, a preconceived opinion touching one of those problems, the solution of which is the ultimate object of the science; and in regard to which, therefore, ethnologists are especially bound to keep their minds open and their judgments freely balanced.

Ethnology, as thus defined, is a branch of ANTHROPOLOGY, the great science which unravels the complexities of human structure; traces out the relations of man to other animals; studies all that is especially human in the mode in which man's complex functions are performed; and searches after the conditions which have determined his presence in the world. And anthropology is a section of ZOOLOGY, which again is the animal half of BIOLOGY—the science of life and living things.

Such is the position of ethnology, such are the objects of the ethnologist. The paths or methods, by following which he may hope to reach his goal, are diverse. He may work at man from the point of view of the pure zoologist, and investigate the anatomical and physiological peculiarities of Negroes, Australians, or Mongolians, just as he would inquire into those of pointers, terriers, and turnspits,—"persistent modifications" of man's almost universal companion. Or he may seek aid from researches into the most human manifestation of humanity—Language; and assuming that what is true of speech is true of the speaker —a hypothesis as questionable in science as it is in ordinary life—he may apply to mankind themselves the conclusions drawn from a searching analysis of their words and grammatical forms.

Or, the ethnologist may turn to the study of the practical life of men; and relying upon 'the inherent conservatism and small inventiveness of untutored mankind,' he may hope to discover in manners and customs, or in weapons, dwellings, and other handiwork, a clue to the origin of 'the resemblances and differences of nations.' Or, he may resort to that kind of evidence which is yielded by History proper, and consists of the beliefs of men concerning past events, embodied in traditional, or in written, testimony. Or, when that thread breaks, Archæology, which is the interpretation of the unrecorded remains of man's works, belonging to the epoch since the world has reached its present condition, may still guide him. And, when even the dim light of archæology fades, there yet remains Palæontology, which, in these latter years, has brought to daylight once more the exuvia of ancient populations, whose world was not our world, who have been buried in river beds immemorially dry, or carried by the rush of waters into caves, inaccessible to inundation since the dawn of tradition.

Along each, or all, of these paths the ethnologist may press towards his goal; but they are not equally straight, or sure, or easy to tread. The way of palæontology has but just been laid open to us. Archæological and historical investigations are of great value for all those peoples whose ancient state has differed widely from their present condition, and who have the good or evil fortune to possess a history. But on taking a broad survey of the world, it is astonishing how few nations present either condition. Respecting five-sixths of the persistent modifications of mankind, history and archæology are absolutely silent. For half the rest, they might as well be silent for anything that is to be made of their testimony. And, finally, when the question arises as to what was the condition of mankind more than a paltry two or three thousand years ago, history and archæology are, for the most part, mere dumb dogs. What light does either of these branches of knowledge throw on the past of the man of the New World, if we except the Central Americans and the Peruvians; on that of the Africans, save those of the Valley of the Nile and a fringe of the Mediterranean; on that of all the Polynesian, Australian, and central Asiatic peoples, the former of whom probably, and the last certainly, were, at the dawn of history, substantially what they are now? While thankfully accepting what history has to give him, therefore, the ethnologist must not look for too much from her.

Is more to be expected from inquiries into the customs and handicrafts of man? It is to be feared not. In reasoning from identity of custom to identity of stock the difficulty always obtrudes itself, that the minds of men being everywhere similar, differing in quality and quantity but not in kind of faculty, like circumstances must tend to produce like contrivances; at any rate, so long as the need to be met and conquered is of a very simple kind. That two nations use calabashes or shells for drinking-vessels, or that they employ spears, or clubs, or swords and axes of stone and metal as weapons and implements, cannot be regarded as evidence that these two nations had a common origin, or even that intercommunication ever took place between them; seeing that the convenience of using calabashes or shells for such purposes, and the advantage of poking an enemy with a sharp stick, or hitting him with a heavy one, must be early forced by nature upon the mind of even the stupidest savage. And when he had found out the use of a stick, he would need no prompting to discover the value of a chipped or whetted stone, or of an angular piece of native metal, for the same object. On the other hand, it may be doubted, whether the chances are not greatly against independent peoples arriving at the manufacture of a boomerang, or of a bow; which last, if one comes to think of it, is a rather complicated apparatus; and the tracing of the distribution of inventions as complex as these, and of such strange customs as betel-chewing and tobacco-smoking, may afford valuable ethnological hints.

Since the time of Leibnitz, and guided by such men as Humboldt, Abel Remusat, and Klaproth, Philology has taken far higher ground. Thus Prichard affirms that "the history of nations, termed Ethnology, must be mainly founded on the relations of their languages."

An eminent living philologer, August Schleicher, in a recent essay, puts forward the claims of his science still more forcibly :—

"If, however, language is the human κατ' ἐξοχήν, the suggestion arises whether it should not form the basis of any scientific systematic arrangement of mankind; whether the foundation of the natural classification of the genus Homo has not been discovered in it.

"How little constant are cranial peculiarities and other so-called race characters! Language, on the other hand, is always a perfectly constant diagnostic. A German may occasionally compete in hair and prognathism with a negro, but a negro language will never be his mother tongue. Of how little importance for mankind the so-called race characters are, is shown by the fact that speakers of languages belonging to one and the same linguistic family may exhibit the peculiarities of various races. Thus the settled Osmanli Turk exhibits Caucasian characters, whilst other so-called Tartaric Turks exemplify the Mongol type. On the other hand, the Magyar and the Basque do not depart in any essential physical peculiarity from the Indo-Germans, whilst the Magyar, Basque, and Indo-Germanic tongues are widely different. Apart from their inconstancy, again, the so-called race characters can hardly yield a scientifically natural system. Languages, on the other hand, readily fall into a natural arrangement, like that of which other vital products are susceptible, especially when viewed from their morphological side. . . . The externally visible structure of the cerebral and facial skeletons, and of the body generally, is less important than that no less material but infinitely more delicate corporeal structure, the function of which is speech. I conceive, therefore, that the natural classification of languages, is also the natural classification of mankind. > With language, moreover, all the higher manifestations of man's vital activity are closely interwoven, so that these receive due recognition in and by that of speech." [1]

Without the least desire to depreciate the value of philology as an adjuvant to ethnology, I must venture to doubt, with Rudolphi, Desmoulins, Crawfurd, and others, its title to the leading position claimed for it by the writers whom I have just quoted. On the contrary, it seems to me obvious that, though, in the absence of any evidence to the contrary, unity of languages may afford a certain presumption in favour of the unity of stock of the peoples speaking those languages, it cannot be held to prove that unity of stock, unless philologers are prepared to demonstrate, that no nation can lose its language and acquire that of a distinct nation, without a change of blood corresponding with the change of language. Desmoulins long ago put this argument exceedingly well :—

"Let us imagine the recurrence of one of those slow, or sudden, political revolutions, or say of those secular changes which among different people and at different epochs have annihilated historical monuments and even extinguished tradition In that case, the evidence, now so clear, that the negroes of Hayti were slaves imported by a French colony, who, by the

---

[1] August Schleicher. *Ueber die Bedeutung der Sprache für die Naturgeschichte des Menschen*, pp. 16—18. Weimar, 1858.

very effect of the subordination involved in slavery lost their own diverse languages and adopted that of their masters, would vanish. And metaphysical philosophers, observing the identity of Haytian French with that spoken on the shores of the Seine and the Loire, would argue that the men of St. Domingo with woolly heads, black and oily skins, small calves, and slightly bent knees, are of the same race, descended from the same parental stock, as the Frenchmen with silky brown, chestnut, or fair hair, and white skins. For they would say, their languages are more similar than French is to German or Spanish."[1]

It must not be imagined that the case put by Desmoulins is a merely hypothetical one. Events precisely similar to the transport of a body of Africans to the West India Islands, indeed, cannot have happened among uncivilised races, but similar results have followed the importation of bodies of conquerors among an enslaved people over and over again. There is hardly a country in Europe in which two or more nations speaking widely different tongues have not become intermixed; and there is hardly a language of Europe of which we have any right to think that its structure affords a just indication of the amount of that intermixture.

As Dr. Latham has well said :—

"It is certain that the language of England is of Anglo-Saxon origin, and that the remains of the original Keltic are unimportant. It is by no means so certain that the blood of Englishmen is equally Germanic. A vast amount of Kelticism,

---

[1] Desmoulins, *Histoire Naturelle des Races Humaines*, p. 345, 1826.

not found in our tongue, very probably exists in our pedigrees. The ethnology of France is still more complicated. Many writers make the Parisian a Roman on the strength of his language; whilst others make him a Kelt on the strength of certain moral characteristics, combined with the previous Kelticism of the original Gauls. Spanish and Portuguese, as languages, are derivations from the Latin; Spain and Portugal, as countries, are Iberic, Latin, Gothic, and Arab, in different proportions. Italian is modern Latin all the world over; yet surely there must be much Keltic blood in Lombardy, and much Etruscan intermixture in Tuscany.

"In the ninth century every man between the Elbe and the Niemen spoke some Slavonic dialect; they now nearly all speak German. Surely the blood is less exclusively Gothic than the speech."[1]

In other words, what philologer, if he had nothing but the vocabulary and grammar of the French and English languages to guide him, would dream of the real causes of the unlikeness of a Norman to a Provençal, of an Orcadian to a Cornishman? How readily might he be led to suppose that the different climatal conditions to which these speakers of one tongue have so long been exposed, have caused their physical differences; and how little would he suspect that these are due (as we happen to know they are) to wide differences of blood.

Few take duly into account the evidence which exists as to the ease with which unlettered savages gain or lose a language. Captain Erskine, in his interesting "Journal of a Cruise among the Islands of the Western Pacific," especially remarks

[1] Latham, *Man and his Migrations*, p. 171.

upon the "avidity with which the inhabitants of the polyglot islands of Melanesia, from New Caledonia to the Solomon Islands, adopt the improvements of a more perfect language than their own, which different causes and accidental communication still continue to bring to them;" and he adds that "among the Melanesian islands scarcely one was found by us which did not possess, in some cases still imperfectly, the decimal system of numeration in addition to their own, in which they reckon only to five."

Yet how much philological reasoning in favour of the affinity or diversity of two distinct peoples has been based on the mere comparison of numerals!

But the most instructive example of the fallacy which may attach to merely philological reasonings, is that afforded by the Feejeans, who are, physically, so intimately connected with the adjacent Negritos of New Caledonia, &c., that no one can doubt to what stock they belong, and who yet, in the form and substance of their language, are Polynesian. The case is as remarkable as if the Canary Islands should have been found to be inhabited by negroes speaking Arabic, or some other clearly Semitic dialect, as their mother tongue. As it happens, the physical peculiarities of the Feejeans are so striking, and the conditions under which they live are so similar to those of the Polynesians, that no one has ventured to suggest that they are merely modified Polynesians—a suggestion which could otherwise certainly have been made. But if languages may be thus transferred from one stock to another, without any corresponding intermixture of blood, what ethnological value has philology?—what security does unity of language afford us that the speakers of that language may not have sprung from two, or three, or a dozen, distinct sources?

Thus we come, at last, to the purely zoological method, from which it is not unnatural to expect more than from any other, seeing that, after all, the problems of ethnology are simply those which are presented to the zoologist by every widely distributed animal he studies. The father of modern zoology seems to have had no doubt upon this point. At the twenty-eighth page of the standard twelfth edition of the "Systema Naturæ," in fact, we find:—

I. PRIMATES.

Dentes primores incisores: superiores IV. paralleli, mammæ pectorales II.

1. HOMO.	Nosce te ipsum.
Sapiens.	1. H. diurnus: *varians cultura, loco.*
Ferus.	Tetrapus, mutus, hirsutus.
. . . . .	. . . .
Americanus a.	Rufus, cholericus, rectus—*Pilis* nigris, rectis, crassis—*Naribus* patulis—*Facie* ephelitica: *Mento* subimberbi.
	Pertinax, contentus, liber. *Pingit* se lineis dædaleis rubris.
	Regitur Consuetudine.

Europæus β. Albus sanguineus torosus. *Pilis* flavescentibus, prolixis.
    Oculis cœruleis.
    Levis, argutus, inventor. *Tegitur* Vestimentis arctis. *Regitur* Ritibus.

Asiaticus γ. Luridus, melancholicus, rigidus. *Pilis* nigricantibus. *Oculis* fuscis. *Severus*, fastuosus, avarus. *Tegitur* Indumentis laxis.
    Regitur Opinionibus.

Afer δ. Niger, phlegmaticus, laxus. *Pilis* atris, contortuplicatis. *Cute* holosericea. *Naso* simo. *Labiis* tumidis. *Feminis* sinus pudoris.
    Mammæ lactantes prolixæ.
    Vafer, segnis, negligens. *Ungit* se pingui.
    Regitur Arbitrio.

Monstrosus ε. Solo (a) et arte (b c) variat. :
    a. *Alpini* parvi, agiles, timidi.
    Patagonici magni, segnes.
    b. *Monorchides* ut minus fertiles : Hottentotti.
    Junceæ puellæ, abdomine attenuato : Europææ.
    c. *Macrocephali* capiti conico : Chinenses.
    Plagiocephali capite antice compresso : Canadenses.

Turn a few pages further on in the same volume, and there appears, with a fine impartiality in the distribution of capitals and subdivisional headings :—

III. FERÆ.

Dentes primores superiores sex, acutiusculi. Canini solitarii.

12. CANIS.   *Dentes primores* superiores VI. : laterales longiores distantes : intermedii lobati. Inferiores VI. : laterales lobati.
    Laniarii solitarii, incurvati.
    Molares VI. s. VII. (pluresve quam in reliquis.)

familiaris	1.	C. cauda (sinistrorsum) recurvata. . . . . . .
domesticus	α.	auriculis erectis, cauda subtus lanata.
sagax	β.	auriculis pendulis, digito spurio ad tibias posticas.
grajus	γ.	magnitudine lupi, trunco curvato, rostro attenuato, &c. &c.

Linnæus' definition of what he considers to be mere varieties of the species Man are, it will be observed, as completely free from any illusion to linguistic peculiarities as those brief and pregnant sentences in which he sketches the characters of the varieties of the species Dog. "Pilis nigris, naribus patulis" may be set against "auriculis erectis, cauda subtus lanata;" while the remarks on the morals and manners of the human subject seem as if they were thrown in merely by way of makeweight.

Buffon, Blumenbach (the founder of ethnology as a special science), Rudolphi, Bory de St. Vincent, Desmoulins, Cuvier, Retzius, indeed I may say all the naturalists proper, have dealt with man from a no less completely zoological point of view; while, as might have been expected, those who have been least naturalists, and most linguists, have most neglected the zoological method, the neglect culminating in those who have been altogether devoid of acquaintance with anatomy.

Prichard's proposition, that language is more persistent than physical characters, is one which has never been proved, and indeed admits of no proof, seeing that the records of language do not extend so far as those of physical characters. But, until the superior tenacity of linguistic over physical peculiarities is shown, and until the abundant evidence which exists, that the language of a people may change without corresponding physical change in that people, is shown to be valueless, it is plain that the zoological court of appeal is the highest for the ethnologist, and that no evidence can be set against that derived from physical characters.

What, then, will a new survey of mankind from the Linnean point of view teach us?

The great antipodal block of land we call Australia has, speaking roughly, the form of a vast quadrangle, 2,000 miles on the side, and extends from the hottest tropical, to the middle of the temperate, zone. Setting aside the foreign colonists introduced within the last century, it is inhabited by people no less remarkable for the uniformity, than for the singularity, of their physical characters and social state. For the most part of fair stature, erect and well built, except for an unusual slenderness of the lower limbs, the AUSTRALIANS have dark, usually chocolate-coloured skins; fine dark wavy hair; dark eyes, overhung by beetle brows; coarse, projecting jaws; broad and dilated, but not especially flattened, noses, and lips which, though prominent, are eminently flexible.

The skulls of these people are always long and narrow, with a smaller development of the frontal sinuses than usually corresponds with such largely developed brow ridges. An Australian skull of a round form, or one the transverse diameter of which exceeds eight-tenths of its length, has never been seen. These people, in a word, are eminently "dolichocephalic," or long-headed; but, with this one limitation, their crania present considerable variations, some being comparatively high and arched, while others are more remarkably depressed than almost any other human skulls. The female pelvis differs comparatively little from the European; but in the pelves of male Australians which I have examined, the antero-posterior and transverse diameters approach equality more nearly than is the case in Europeans.

No Australian tribe has ever been known to cultivate the ground,[1] to use metals, pottery, or any kind of textile fabric. They rarely construct huts. Their means of navigation are limited to rafts or canoes, made of sheets of bark. Clothing, except skin cloaks for protection from cold, is a superfluity with which they dispense; and though they have some singular weapons, almost peculiar

[[1] At Cape York we found that the natives had learned from their Papuan neighbours to grow a little coarse tobacco; and, elsewhere, yams are said to be grown, but hardly cultivated. Plaiting, basket-making, and netting are practised.—1894.]

to themselves, they are wholly unacquainted with bows and arrows.

It is but a step, as it were, across Bass's Straits to Tasmania. Neither climate nor the characteristic forms of vegetable or animal life change largely on the south side of the Straits, but the early voyagers found Man singularly different from him on the north side. The skin of the Tasmanian was dark, though he lived between parallels of latitude corresponding with those of middle Europe in our own hemisphere; his jaws projected, his head was long and narrow; his civilization was about on a footing with that of the Australian, if not lower, for I cannot discover that the Tasmanian understood the use of the throwing-stick. But he differed from the Australian in his woolly, negro-like hair; whence the name of NEGRITO, which has been applied to him and his congeners.

Such Negritos—differing more or less from the Tasmanian but agreeing with him in dark skin and woolly hair—occupy New Caledonia, the New Hebrides, the Louisiade Archipelago; and stretching to the Papuan Islands, and for a doubtful extent beyond them to the north and west, form a sort of belt, or zone, of Negrito population, interposed between the Australians on the west and the inhabitants of the great majority of the Pacific islands on the east.

The cranial characters of the Negritos vary considerably more than those of their skin and hair, the most notable circumstance being the strong Australian aspect which distinguishes many Negrito skulls, while others tend rather towards forms common in the Polynesian islands.

In civilization, New Caledonia exhibits an advance upon Tasmania, and, farther north, there is a still greater improvement. But the bows and arrows, the perched houses, the outrigger canoes, the habits of betel-chewing and of kawa-drinking, which abound more or less among the northern Negritos, are probably to be regarded not as the products of an indigenous civilization, but merely as indications of the extent to which foreign influences have modified the primitive social state of these people.

From Tasmania or New Caledonia, to New Zealand or Tongataboo, is again but a brief voyage: but it brings about a still more notable change in the aspect of the indigenous population than that effected by the passage of Bass's Straits. Instead of being chocolate-coloured people, the Maories and Tongans are light brown; instead of woolly, they have straight, or wavy, black hair. And if from New Zealand, we travel some 5,000 miles east to Easter Island; and from Easter Island, for as great a distance north-west, to the Sandwich Islands; and thence 7,000 miles, westward and southward, to Sumatra; and even across the Indian Ocean, into the interior of Madagascar, we shall everywhere meet with people whose hair is straight or wavy, and whose skins exhibit various shades of brown. These are the Polynesians, Micronesians, Indonesians, whom Latham has grouped together under the common title of AMPHINESIANS.

The cranial characters of these people, as of the Negritos, are less constant than those of their skin and hair. The Maori has a long skull; the Sandwich Islander a broad skull. Some, like these, have strong brow ridges; others like the Dayaks and many Polynesians, have hardly any nasal indentation. It is only in the westernmost parts of their area that the Amphinesian nations know anything about bows and arrows as weapons, or are acquainted with the use of metals or with pottery. Everywhere they cultivate the ground, construct houses, and skilfully build and manage outrigger, or double, canoes; while, almost everywhere, they use some kind of fabric for clothing.

Between Easter Island, or the Sandwich Islands, and any part of the American coast is a much wider interval than that between Tasmania and New Zealand, but the ethnological interval between the American and the Polynesian is less than that between either of the previously named stocks.

The typical AMERICAN has straight black hair and dark eyes, his skin exhibiting various shades of reddish or yellowish brown, sometimes inclining to olive. The face is broad and scantily bearded;

the skull wide and high. Such people extend from Patagonia to Mexico, and much farther north along the west coast. In the main a race of hunters, they had nevertheless, at the time of the discovery of the Americas, attained a remarkable degree of civilization in some localities. They had domesticated ruminants, and not only practised agriculture, but had learned the value of irrigation. They manufactured textile fabrics, were masters of the potter's art, and knew how to erect massive buildings of stone. They understood the working of the precious, though not of the useful, metals;[1] and had even attained to a rude kind of hieroglyphic, or picture, writing. The Americans not only employ the bow and arrow, but, like some Amphinesians, the blow-pipe, as offensive weapons: but I am not aware that the outrigger canoe has ever been observed among them.

I have reason to suspect that some of the Fuegian tribes differ cranially from the typical Americans;[2] and the Northern and Eastern American tribes have longer skulls than their Southern compatriots. But the ESQUIMAUX, who roam on the desolate and ice-bound coast of Arctic America, certainly present us with a new stock. The Esquimaux (among whom the Greenlanders are included), in fact, though they share the straight

[1 With the exception of copper and bronze.—1894.]
[2 A suspicion subsequently verified. See a memoir on American Skulls, *Journal of Anatomy and Physiology*. Vol. 16. —1894.]

black hair of the proper Americans, are generally a duller complexioned, shorter, and a more squat people, and they have still more prominent cheek-bones. But the circumstance which most completely separates them from the typical Americans, is the form of their skulls, which instead of being broad, high, and truncated behind, are eminently long, usually low, and prolonged backwards. These Hyperborean people clothe themselves in skins, know nothing of pottery, and hardly anything of metals. Dependent for existence upon the produce of the chase, the seal and the whale are to them what the cocoa-nut tree and the plantain are to the savages of more genial climates. Not only are those animals meat and raiment, but they are canoes, sledges, weapons, tools, windows, and fire; while they support the dog, who is the indispensable ally and beast of burden of the Esquimaux.

It is admitted that the Tchuktchi, on the eastern side of Behring's Straits, are, in all essential respects, Esquimaux; and I do not know that there is any satisfactory evidence to show that the Tunguses and Samoiedes do not essentially share the same physical characters. Southward, there are indications of Esquimaux characters among the Japanese, and it is possible that their influence may be traced yet further.

However this may be, Eastern Asia, from Mantchouria to Siam, Thibet, and Northern Hindostan, is continuously inhabited by men, usually of short stature, with skins varying in colour from yellow to olive; with broad cheek-bones and faces that, owing to the insignificance of the nose, are exceedingly flat; and with small, obliquely-set[1] black eyes and straight black hair, which sometimes attains a very great length upon the scalp, but is always scanty upon the face and body. The skull, never much elongated, is, generally, remarkably broad and rounded, with hardly any nasal depression, and but slight, if any, projection of the jaws. Many of these people, for whom the old name of MONGOLIANS may be retained, are nomades; others, as the Chinese, have attained a remarkable and apparently indigenous civilization, only surpassed by that of Europe.

At the north-western extremity of Europe the Lapps repeat the characters of the Eastern Asiatics. Between these extreme points, the Mongolian stock is not continuous, but is represented by a chain of more or less isolated tribes, who pass under the name of Calmucks and Tartars, and form Mongolian islands, as it were, in the midst of an ocean of other people.

The waves of this ocean are the nations for whom, in order to avoid the endless confusion produced by our present half-physical, half-philo-

[1 The obliquity, it must be recollected, is not in the position of the eyeball but arises from the arrangement of the skin in the neighbourhood of the eyelids.—1894]

logical classification, I shall use a new name—XANTHOCHROI—indicating that they are "yellow" haired and "pale" in complexion. The Chinese historians of the Han dynasty, writing in the third century before our era, describe, with much minuteness, certain numerous and powerful barbarians with "yellow hair, green eyes, and prominent noses," who, the black-haired, skew-eyed, and flat-nosed annalists remark in passing, are "just like the apes from whom they are descended." These people held, in force, the upper waters of the Yenisei, and thence under various names stretched southward to Thibet and Kashgar. Fair-haired and blue-eyed northern enemies were no less known to the ancient Hindoos, to the Persians, and to the Egyptians, on the south and west of the great central Asiatic area; while the testimony of all European antiquity is to the effect that, before and since the period in question, there lay beyond the Danube, the Rhine, and the Seine, a vast and dangerous yellow or red haired, fair-skinned, blue-eyed population. Whether the disturbers of the marches of the Roman Empire were called Gauls or Germans, Goths, Alans, or Scythians, one thing seems certain, that until the invasion of the Huns, they were largely tall, fair, blue-eyed men.

If any one should think fit to assume that, in the year 100 B.C., there was one continuous Xanthochroic population from the Rhine to the

Yenisei, and from the Ural mountains to the Hindoo Koosh, I know not that any evidence exists by which that position could be upset, while the existing state of things is rather in its favour than otherwise. For the Scandinavians, the Germans, the Slavonian and the Finnish tribes, to a great extent; some of the inhabitants of Greece, many Turks, some Kirghis, and some Mantchous, the Ossetes in the Caucasus, the Siahposh, the Rohillas, are at the present day fair, yellow or red haired, and blue-eyed; and the interpolation of tribes of Mongolian hair and complexion, as far west as the Caspian Steppes and the Crimea, might justly be accounted for by those subsequent westward irruptions of the Mongolian stock, of which history furnishes abundant testimony. The furthermost limit of the Xanthochroi north-westward is Iceland and the British Isles; south-westward, they are traceable at intervals through Syria and the Berber country, ending in the Canary Islands. The cranial characters of the Xanthochroi are not, at present, strictly definable. The Scandinavians are certainly long-headed; but many Germans, the Swiss so far as they are Germanized, the Slavonians, the Fins, and the Turks, are short-headed. What were the cranial characters of the ancient "U-suns" and "Tinglings" of the valley of the Yenisei is unknown.

West and south of the area occupied by the chief mass of the Xanthochroi, and north of the

Sahara, is a broad belt of land, shaped like a ➤.
Between the forks of the Y lies the Mediterranean;
the stem of it is Arabia. The stem is bathed by
the Indian Ocean, the western ends of the forks
by the Atlantic. The majority of the people in-
habiting the area thus roughly defined have, like
the Xanthochroi, prominent noses, pale skins and
wavy hair, with abundant beards; but, unlike them,
the hair is black or dark and the eyes usually so.
They may thence be called the MELANOCHROI.
Such people are found in the British Islands, in
Western and Southern Gaul, in Spain, in Italy
south of the Po, in parts of Greece, in Syria and
Arabia, stretching as far northward and eastward
as the Caucasus and Persia. They are the chief
inhabitants of Africa north of the Sahara, and, like
the Xanthochroi, they end in the Canary Islands.
They are known as Kelts, Iberians, Etruscans,
Romans, Pelasgians, Berbers, Semites. The
majority of them are long-headed, and of smaller
stature than the Xanthochroi.[1] It is needless
to remark upon the civilization of these two
great stocks. With them has originated everything that is highest in science, in art, in law,
in politics, and in mechanical inventions. In their
hands, at the present moment, lies the order of the
social world, and to them its progress is committed.

South of the Atlas, and of the Great Desert,

[[1] See the Essay on the Aryan Question, in this volume, for some qualifications of these statements necessitated by further knowledge. 1894.]

Middle Africa exhibits a new type of humanity in the NEGRO, with his dark skin, woolly hair, projecting jaws, and thick lips. As a rule, the skull of the Negro is remarkably long; it rarely approaches the broad type, and never exhibits the roundness of the Mongolian. A cultivator of the ground, and dwelling in villages; a maker of pottery, and a worker in the useful as well as the ornamental metals; employing the bow and arrow as well as the spear, the typical negro stands high in point of civilization above the Australian.

Resembling the Negroes in cranial characters, the BUSHMEN of South Africa differ from them in their yellowish brown skins, their tufted hair, their remarkably small stature, and their tendency to fatty and other integumentary outgrowths; nor is the wonderful click with which their speech is interspersed to be overlooked in enumerating the physical characteristics of this strange people.

The so-called "Dravidian" populations of Southern Hindostan lead us back, physically as well as geographically, towards the Australians;[1]

[[1] Of the affinities of these stocks I think there can be no doubt. I was formerly inclined to believe that the ancient Egyptian was the highest term in an ascending series: Australian—Dravidian—Egyptian of allied stocks. And I believe still that there is a good deal to be said for that hypothesis. One of the most interesting problems at present is the relation of the præ-semitic population of Babylonia to the Dravidians, on the one hand, and the Old Egyptian on the other. Only one point appears to me to be quite clear, if the statues of Tell Loh represent these people; that there is not a trace of Mongolian affinity about them.—1894.]

while the diminutive MINCOPIES of the Andaman Islands lie midway between the Negro and Negrito races, and, as Mr. Busk has pointed out, occasionally present the rare combination of brachycephaly, or short-headedness, with woolly hair.

In the preceding progress along the outskirts of the habitable world, eleven readily distinguishable stocks, or persistent modifications, of mankind, have been recognized. I have purposely omitted such people as the Abyssinians and the Hindoos of the valleys of the Ganges and Indus, who there is every reason to believe result from the intermixture of distinct stocks. Perhaps I ought for like reasons, to have ignored the Mincopies. But I do not pretend that my enumeration is complete or, in any sense, perfect. It is enough for my purpose if it be admitted (and I think it cannot be denied) that those which I have mentioned exist, are well marked, and occupy the greater part of the habitable globe.

In attempting to classify these persistent modifications after the manner of naturalists, the first circumstance that attracts one's attention is the broad contrast between the people with straight and wavy hair, and those with crisp, woolly, or tufted hair. Bory de St. Vincent, noting this fundamental distinction, divided mankind accordingly into the two primary groups of *Leiotrichi* and *Ulotrichi*,—terms which are open to criticism, but which I adopt in the accompanying table, because they have been used. It is better for science to accept a faulty name which has the merit of existence, than to burthen it with a faultless newly invented one.

Under each of these divisions are two columns, one for the Brachycephali, or short heads, and one for the Dolichocephali,[1] or long heads. Again, each column is subdivided transversely into four compartments, one for the "leucous," people with fair complexions and yellow or red hair; one for the "leucomelanous," with dark hair and pale skins; one for the "xanthomelanous," with black hair and yellow, brown, or olive skins; and one for the "melanous," with black hair and dark brown or blackish skins.

	LEIOTRICHI.		ULOTRICHI.	
	Dolichocephali.	Brachycephali.	Dolichocephali.	Brachycephali.
Leucous.				
	.... Xanthochroi			
Leucomelanous.				
	.... Melanochroi			
Xanthomelanous.				
	Esquimaux.	Mongolians.	*Bushmen.*	
	Amphinesians.			
	Americans.			
Melanous.				
	Australians.		Negroes.	*Mincopies* (?)
			Negritos.	

⁎⁎ *The names of the stocks known only since the fifteenth century are put into italics. If the "Skrälings" of the Norse discoverers of America were Esquimaux, Europeans became acquainted with the latter six or seven centuries earlier.*

[1] Skulls, the transverse diameter of which is more than eight-

It is curious to observe that almost all the woolly-haired people are also long-headed; while among the straight-haired nations broad heads preponderate, and only two stocks, the Esquimaux and the Australians, are exclusively long-headed.

One of the acutest and most original of ethnologists, Desmoulins, originated the idea, which has subsequently been fully developed by Agassiz, that the distribution of the persistent modifications of man is governed by the same laws as that of other animals, and that both fall into the same great distributional provinces. Thus, Australia; America, south of Mexico; the Arctic regions; Europe, Syria, Arabia, and North Africa, taken together, are each regions eminently characterised by the nature of their animal and vegetable populations, and each, as we have seen, has its peculiar and characteristic form of man. But it may be doubted whether the parallel thus drawn will hold good strictly, and in all cases. The Tasmanian Fauna and Flora are essentially Australian, and the like is true, to a less extent, of many, if not of all, the Papuan islands; but the Negritos who inhabit these islands are strikingly different from the Australians. Again, the differences between the Mongolians and the Xanthochroi are out of all proportion greater than those

tenths the long diameter, are short; those which have the transverse diameter less than eight-tenths the longitudinal, are long.

between the Faunæ and Floræ of Central and Eastern Asia. But whatever the difficulties in the way of the detailed application of this comparison of the distribution of men with that of animals, it is well worthy of being borne in mind, and carried as far as it will go.

Apart from all speculation, a very curious fact regarding the distribution of the persistent modifications of mankind becomes apparent on inspecting an Ethnological chart, projected in such a manner that the Pacific Ocean occupies its centre. Such a chart exhibits an Australian area occupied by dark smooth-haired people, separated by an incomplete inner zone of dark woolly-haired Negritos and Negroes, from an outer zone of comparatively pale and smooth-haired men, occupying the Americas, and nearly all Asia[1] and North Africa.[2]

Such is a brief sketch of the characters and distribution of the persistent modifications, or stocks, of mankind at the present day. If we seek for direct evidence of how long this state of things has lasted, we shall find little enough, and that little far from satisfactory. Of the eleven different stocks enumerated, seven have been known to us for less than 400 years; and of these seven not one possessed a fragment of written history at the

[1 Hindostan excepted.—1894.]
[2 Egypt excepted.—1894.]

time it came into contact with European civilization. The other four—the Negroes, Mongolians, Xanthochroi, and Melanochroi—have always existed in some of the localities in which they are now found, nor do the negroes ever seem to have voluntarily travelled beyond the limits of their present area. But ancient history is in a great measure the record of the mutual encroachments of the other three stocks.

On the whole, however, it is wonderful how little change has been effected by these mutual invasions and intermixtures. As at the present time, so at the dawn of history, the Melanochroi fringed the Atlantic and the Mediterranean; the Xanthochroi occupied most of Central and Eastern Europe, and much of Western and Central Asia; while Mongolians held the extreme east of the Old World. So far as history teaches us, the populations of Europe, Asia and Africa were, twenty[1] centuries ago, just what they are now, in their broad features and general distribution.

The evidence yielded by Archæology is not very definite, but so far as it goes, it is to much the same effect. The mound builders of Central America seem to have had the characteristic short and broad head of the modern inhabitants of that continent. The tumuli and tombs of Ancient Scandinavia, of pre-Roman Britain, of Gaul, of

[[1] We may now safely say thirty or forty.—1894.]

Switzerland, reveal two types of skull—a broad and a long—of which, in Scandinavia, the broad seems to have belonged to the older stock, while the reverse was probably the case in Britain, and certainly in Switzerland. It has been assumed that the broad-skulled people of ancient Scandinavia were Lapps; but there is no proof of the fact, and they may have been, like the broad-skulled Swiss and Germans, Xanthochroi. One of the greatest of ethnological difficulties is to know where the modern Swedes, Norsemen, and Saxons got their long heads, as all their neighbours, Fins, Lapps, Slavonians, and South Germans, are broad-headed. Again, who were the small-handed [1] long-headed people of the "bronze epoch," and what has become of the infusion of their blood among the Xanthochroi?

At present Palæontology yields no safe data to the ethnologist. We know absolutely nothing of the ethnological characters of the men of Abbeville and Hoxne; but must be content with the demonstration, in itself of immense value, that Man existed in Western Europe when its physical condition was widely different from what it is now, and when animals existed, which, though they belong to what is, properly speaking, the present

[[1] Supposed to be small-handed from the small handles of their bronze swords. But I observe in the Assyrian sculptures the same small handles, while the hands are by no means small. How did the Assyrians use their swords? So far as I know thrusting alone is represented.—1894.]

order of things, have long been extinct. Beyond the limits of a fraction of Europe, Palæontology tells us nothing of man or of his works.

To sum up our knowledge of the ethnological past of man; so far as the light is bright, it shows him substantially as he is now; and, when it grows dim, it permits us to see no sign that he was other than he is now.

It is a general belief that men of different stocks differ as much physiologically as they do morphologically; but it is very hard to prove, in any particular case, how much of a supposed national characteristic is due to inherent physiological peculiarities, and how much to the influence of circumstances. There is much evidence to show, however, that some stocks enjoy a partial or complete immunity from diseases which destroy, or decimate, others. Thus there seems good ground for the belief that Negroes are remarkably exempt from yellow fever; and that, among Europeans, the melanochroic people are less obnoxious to its ravages than the xanthochroic. But many writers, not content with physiological differences of this kind, undertake to prove the existence of others of far greater moment; and, indeed, to show that certain stocks of mankind exhibit, more or less distinctly, the physiological characters of true species. Unions between these stocks, and still more between the half-breeds arising from their mixture, are affirmed to be either infertile, or less fertile than those which take place between males and females of either stock under the same circumstances. Some go so far as to assert that no mixed breeds of mankind can maintain themselves without the assisttance of one or other of the parent stocks, and that, consequently, they must inevitably be obliterated in the long run.

Here, again, it is exceedingly difficult to obtain trustworthy evidence and to free the effects of the pure physiological experiment from adventitious influences. The only trial which, by a strange chance, was kept clear of all such influences —the only instance in which two distinct stocks of mankind were crossed, and their progeny intermarried without any admixture from without— is the famous case of the Pitcairn Islanders, who were the progeny of Bligh's English sailors by Tahitian women. The results of this experiment, as everybody knows, are dead against those who maintain the doctrine of human hybridity, seeing that the Pitcairn Islanders, even though they necessarily contracted consanguineous marriages, throve and multiplied exceedingly.

But those who are disposed to believe in this doctrine should study the evidence brought forward in its support by M. Broca, its latest and ablest advocate, and compare this evidence with that which the botanists, as represented by a Gaertner, or by a Darwin, think it indispensable to obtain before they will admit the infertility of crosses between two allied kinds of plants. They will then, I think, be satisfied that the doctrine in question rests upon a very unsafe foundation; that the facts adduced in its support are capable of many other interpretations; and, indeed, that from the very nature of the case, demonstrative evidence one way or the other is almost unattainable. *A priori*, I should be disposed to expect a certain amount of infertility between some of the extreme modifications of mankind; and still more between the offsprings of their intermixture. *A posteriori*, I cannot discover any satisfactory proof that such infertility exists.

From the facts of ethnology I now turn to the theories and speculations of ethnologists, which have been devised to explain these facts, and to furnish satisfactory answers to the inquiry —what conditions have determined the existence of the persistent modifications of mankind, and have caused their distribution to be what it is?

These speculations may be grouped under three heads: firstly the Monogenist hypotheses; secondly, those of the Polygenists; and thirdly, that which would result from a simple application of Darwinian principles to mankind.

According to the Monogenists, all mankind have sprung from a single pair, whose multitudinous progeny spread themselves over the world, such as it now is, and became modified into the forms we meet with in the various regions of the earth, by the effect of the climatal and other conditions to which they were subjected.

The advocates of this hypothesis are divisible into several schools. There are those who represent the most numerous, respectable, and would-be orthodox of the public, and are what may be called "Adamites," pure and simple. They believe that Adam was made out of earth somewhere in Asia, about six thousand years ago; that Eve was was modelled from one of his ribs; and that the progeny of these two having been reduced to the eight persons who were landed on the summit of Mount Ararat after an universal deluge, all the nations of the earth have proceeded from these last, have migrated to their present localities, and have become converted into Negroes, Australians, Mongolians, &c., within that time. Five-sixths of the public are taught this Adamitic Monogenism, as if it were an established truth, and believe it. I do not; and I am not acquainted with any man of science, or duly instructed person, who does.

A second school of monogenists, not worthy of much attention, attempts to hold a place midway between the Adamites and a third division, who take up a purely scientific position, and require to be dealt with accordingly. This third division, in fact, numbers in its ranks Linnæus, Buffon,

Blumenbach, Cuvier, Prichard, and many distinguished living ethnologists.

These "Rational Monogenists," or, at any rate, the more modern among them, hold, firstly, that the present condition of the earth has existed for untold ages; secondly, that, at a remote period, beyond the ken of Archbishop Usher, man was created, somewhere between the Caucasus and the Hindoo Koosh; thirdly, that he might have migrated thence to all parts of the inhabited world, seeing that none of them are unattainable from some other inhabited part, by men provided with only such means of transport as savages are known to possess and must have invented; fourthly, that the operation of the existing diversities of climate and other conditions upon people so migrating, is sufficient to account for all the diversities of mankind.

Of the truth of the first of these propositions no competent judge now entertains any doubt. The second is more open to discussion; for, in these latter days, many question the special creation of man: and even if his special creation be granted, there is not a shadow of a reason why he should have been created in Asia rather than anywhere else. Of all the odd myths that have arisen in the scientific world, the "Caucasian mystery," invented quite innocently by Blumenbach, is the oddest. A Georgian woman's skull was the handsomest in his collection. Hence it became his model exemplar of human skulls, from which all others might be regarded as deviations; and out of this, by some strange intellectual hocus-pocus, grew up the notion that the Caucasian man is the prototypic "Adamic" man, and his country the primitive centre of our kind. Perhaps the most curious thing of all is, that the said Georgian skull, after all, is not a skull of average form, but distinctly belongs to the brachycephalic group.

With the third proposition I am quite disposed to agree, though it must be recollected that it is one thing to allow that a given migration is possible, and another to admit there is good reason to believe it has really taken place.

But I can find no sufficient ground for accepting the fourth proposition; and I doubt if it would ever have obtained its general currency except for the circumstance that fair Europeans are very readily tanned and embrowned by the sun. Yet I am not aware that there is a particle of proof that the cutaneous change thus effected can become hereditary, any more than that the enlarged livers, which plague our countrymen in India, can be transmitted; while there is very strong evidence to the contrary. Not only, in fact, are there such cases as those of the English families in Barbadoes, who have remained for six generations unaltered in complexion, but which are open to the objection that they may have received infusions of fresh European blood; but there is the broad fact, that not a single indigenous Negro exists either in the great alluvial plains of tropical South America, or in the exposed islands of the Polynesian Archipelago, or among the populations of equatorial Borneo or Sumatra. No satisfactory explanation of these obvious difficulties has been offered by the advocates of the direct influence of conditions. And as for the more important modifications observed in the structure of the brain, and in the form of the skull, no one has ever pretended to show in what way they can be effected directly by climate.

It is here, in fact, that the strength of the Polygenists, or those who maintain that men primitively arose, not from one, but from many stocks, lies. Show us, they say to the Monogenists, a single case in which the characters of a human stock have been essentially modified without its being demonstrable, or, at least, highly probable, that there has been intermixture of blood with some foreign stock. Bring forward any instance in which a part of the world, formerly inhabited by one stock, is now the dwelling-place of another, and we will prove the change to be the result of migration, or of intermixture, and not of modification of character by climatic influences. Finally, prove to us that the evidence in favour of the specific distinctness of many animals, admitted to be distinct species by all zoologists, is a whit better than that upon which we maintain the specific distinctness of men.

If presenting unanswerable objections to your adversary were the same thing as proving your own case, the Polygenists would be in a fair way towards victory; but, unfortunately, as I have already observed, they have as yet completely failed to adduce satisfactory positive proof of the specific diversity of mankind. Like the Monogenists, the Polygenists are of several sects; some imagine that their assumed species of mankind were created where we find them—the African in Africa, and the Australian in Australia, along with the other animals of their distributional province; others conceive that each species of man has resulted from the modification of some antecedent species of ape—the American from the broad-nosed Simians of the New World, the African from the Troglodytic stock, the Mongolian from the Orangs.

The first hypothesis is hardly likely to win much favour. The whole tendency of modern science is to thrust the origination of things further and further into the background; and the chief philosophical objection to Adam being, not his oneness, but the hypothesis of his special creation; the multiplication of that objection tenfold is, whatever it may look, an increase, instead of a diminution, of the difficulties of the case. And, as to the second alternative, it may safely be affirmed that, even if the differences between men are specific, they are so small, that the assumption of more than one primitive stock for all is altogether superfluous. Surely no one can now be found to assert that any two stocks of mankind differ as much as a chimpanzee and an orang do; still less that they are as unlike as either of these is to any New World Simian!

Lastly, the granting of the Polygenist premises does not, in the slightest degree, necessitate the Polygenist conclusion. Admit that Negroes and Australians, Negritos and Mongols are distinct species, or distinct genera, if you will, and you may yet, with perfect consistency, be the strictest of Monogenists, and even believe in Adam and Eve as the primæval parents of all mankind.

It is to Mr. Darwin we owe this discovery: it is he who, coming forward in the guise of an eclectic philosopher, presents his doctrine as the key to ethnology, and as reconciling and combining all that is good in the Monogenistic and Polygenistic schools. It is true that Mr. Darwin has not, in so many words, applied his views to ethnology; but even he who "runs and reads" the "Origin of Species" can hardly fail to do so; and, furthermore, Mr. Wallace and M. Pouchet have recently treated of ethnological questions from this point of view. Let me, in conclusion, add my own contribution to the same store.

I assume Man to have arisen in the manner which I have discussed elsewhere, and probably, though by no means necessarily, in one locality. Whether he arose singly, or a number of examples appeared contemporaneously, is also an open question for the believer in the production of species by the gradual modification of pre-existing ones. At what epoch of the world's history this took place, again, we have no evidence whatever. It may have been in the older tertiary, or earlier; but what is most important to remember is, that the discoveries of late years have proved that man inhabited Western Europe, at any rate, before the occurrence of those great physical changes which have given Europe its present aspect. And as the same evidence shows that man was the contemporary of animals which are now extinct, it is not too much to assume that his existence dates back at least as far as that of our present Fauna and Flora, or before the epoch of the drift.

But if this be true, it is somewhat startling to reflect upon the prodigious changes which have taken place in the physical geography of this planet since man has been an occupant of it.

During that period the greater part of the British islands, of Central Europe, of Northern Asia, have been submerged beneath the sea and raised up again. So has the great desert of Sahara, which occupies the major part of Northern

Africa.[1] The Caspian and the Aral seas have been one, and their united waters have probably communicated with both the Arctic and the Mediterranean oceans.[2] The greater part of North America has been under water, and has emerged. It is highly probable that a large part of the Malayan Archipelago has sunk, and that its primitive continuity with Asia has been destroyed. Over the great Polynesian area subsidence has taken place to the extent of many thousands of feet—subsidence of so vast a character, in fact, that if a continent like Asia had once occupied the area of the Pacific, the peaks of its mountains would now show not more numerous than the islands of the Polynesian Archipelago.[3]

What lands may have been thickly populated for untold ages, and subsequently have disappeared and left no sign above the waters, it is of course impossible for us to say; but unless we are to make the wholly unjustifiable assumption that no dry land rose elsewhere when our present dry land sank, there must be half-a-dozen Atlantises beneath the waves of the various oceans of the world. But if the regions which have undergone

[1 Later investigations tend to show that only a small part of the Sahara has been submerged.—1894.]

[2 With reference to certain reclamations that have been made à propos of a speculation set forth in the essay on the *Aryan Question* (*infrà*), I draw attention to the fact that this passage was written twenty-nine years ago.—1894.]

[3 The occurrence of this extensive subsidence is disputed.—1894.]

these slow and gradual, but immense alterations, were wholly or in part inhabited before the changes I have indicated began—and it is more probable that they were than that they were not—what a wonderfully efficient "Emigration Board" must have been at work all over the world long before canoes, or even rafts, were invented; and before men were impelled to wander by any desire nobler or stronger than hunger. And as these rude and primitive families were thrust, in the course of long series of generations, from land to land, impelled by encroachments of sea or of marsh, or by severity of summer heat or winter cold, to change their positions, what opportunities must have been offered for the play of natural selection, in preserving one family variation and destroying another!

Suppose, for example, that some families of a horde which had reached a land charged with the seeds of yellow fever, varied in the direction of woolliness of hair and darkness of skin. Then, if it be true that these physical characters are accompanied by comparative or absolute exemptions from that scourge, the inevitable tendency would be to the preservation and multiplication of the darker and woollier families, and the elimination of the whiter and smoother haired. In fact, by the operation of causes precisely similar to those which, in the famous instance cited by Mr. Darwin, have given rise to a race of black pigs in the forests of Louisiana, a negro stock would eventually people the region.[1] Again, how often, by such physical changes, must a stock have been isolated from all others for innumerable generations, and have found ample time for the hereditary hardening of its special peculiarities into the enduring characters of a persistent modification.

Nor, if it be true that the physiological differences of species may be produced by variation and natural selection, as Mr. Darwin supposes, would it be at all astonishing, if, in some of these separated stocks, the process of differentiation should have gone so far as to give rise to the phenomena of hybridity. In the face of the overwhelming evidence in favour of the unity of the origin of mankind afforded by anatomical considerations, satisfactory proof of the existence of any degree of sterility in the unions of members of two of the "persistent modifications" of mankind, might well be appealed to by Mr. Darwin as crucial evidence of the truth of his views regarding the origin of species in general.

[[1] Mr. Pearson, in his very interesting work *On National Life and Character*, justly dwells upon the obstacles to the existence of the white races within the Tropics. There is, however, this point to be considered, that the fevers to which the white men succumb are probably caused by microbes; and that modern therapeutic science is daily teaching us more and more about the ways of obtaining immunity from or alleviating these attacks. What would become of black competition if fever "vaccination" proved effectual?—1894.]

V

ON SOME FIXED POINTS IN BRITISH ETHNOLOGY

[1871]

IN view of the many discussions to which the complicated problems offered by the ethnology of the British Islands have given rise, it may be useful to attempt to pick out, from amidst the confused masses of assertion and of inference, those propositions which appear to rest upon a secure foundation, and to state the evidence by which they are supported. Such is the purpose of the present paper.

Some of these well-based propositions relate to the physical characters of the people of Britain and their neighbours; while others concern the languages which they spoke. I shall deal, in the first place, with the physical questions.

I. *Eighteen hundred years ago the population of Britain comprised people of two types of complexion*

—the one fair, and the other dark. The dark people resembled the Aquitani and the Iberians; the fair people were like the Belgic Gauls.

The chief direct evidence of the truth of this proposition is the well-known passage of Tacitus:—

"Ceterum Britanniam qui mortales initio coluerint, indigenæ an advecti, ut inter barbaros, parum compertum. Habitus corporum varii: atque ex eo argumenta: namque rutilæ Caledoniam habitantium comæ, magni artus, Germanicam originem asseverant. Silurum colorati vultus et torti plerumque crines, et posita contra Hispania, Iberos veteres trajecisse, easque sedes occupasse, fidem faciunt. Proximi Gallis et similes sunt; seu durante originis vi, seu procurrentibus in diversa terris, positio cœli corporibus habitum dedit. In universum tamen æstimanti, Gallos vicinum solum occupasse, credibile est; eorum sacra deprehendas, superstitionum persuasione; sermo haud multum diversus."[1]

This passage, it will be observed, contains statements as to facts, and certain conclusions deduced from these facts. The matters of fact asserted are: firstly, that the inhabitants of Britain exhibit much diversity in their physical characters; secondly, that the Caledonians are red-haired and large-limbed, like the Germans; thirdly, that the Silures have curly hair and dark complexions, like the people of Spain; fourthly, that the British people nearest Gaul resemble the "Galli."

Tacitus, therefore, states positively what the Caledonians and Silures were like; but the

[1] Tacitus *Agricola*, c. 11.

interpretation of what he says about the other Britons must depend upon what we learn from other sources as to the characters of these "Galli." Here the testimony of "divus Julius" comes in with great force and appropriateness. Cæsar writes :—

> "Britanniæ pars interior ab iis incolitur, quos natos in insula ipsi memoria proditum dicunt : maritima pars ab iis, qui prædæ ac belli inferendi causa ex Belgio transierant ; qui omnes fere iis nominibus civitatum appellantur quibus orti ex civitatibus eo pervenerunt, et bello inlato ibi permanserunt atque agros colere cœperunt." [1]

From these passages it is obvious that, in the opinion of Cæsar and Tacitus, the southern Britons resembled the northern Gauls, and especially the Belgæ; and the evidence of Strabo is decisive as to the characters in which the two people resembled one another: "The men [of Britain] are taller than the Kelts, with hair less yellow; they are slighter in their persons." [2]

The evidence adduced appears to leave no reasonable ground for doubting that, at the time of the Roman conquest, Britain contained people of two types, the one dark and the other fair complexioned, and that there was a certain difference between the latter in the north and in the south of Britain: the northern folk being, in the judg-

---

[1] *De Bello Gallico*, v. 12.
[2] The Geography of Strabo. Translated by Hamilton and Falconer, v. 5.

ment of Tacitus, or, more properly, according to the information he had received from Agricola and others, more similar to the Germans than the latter. As to the distribution of these stocks, all that is clear is, that the dark people were predominant in certain parts of the west of the southern half of Britain, while the fair stock appears to have furnished the chief elements of the population elsewhere.

No ancient writer troubled himself with measuring skulls, and therefore there is no direct evidence as to the cranial characters of the fair and the dark stocks. The indirect evidence is not very satisfactory. The tumuli of Britain of pre-Roman date have yielded two extremely different forms of skull, the one broad and the other long; and the same variety has been observed in the skulls of the ancient Gauls.[1] The suggestion is obvious that the one form of skull may have been associated with the fair and the other with the dark, complexion. But any conclusion of this kind is at once checked by the reflection that the extremes of long and short-headedness are to be met with among the fair inhabitants of Germany and of Scandinavia at the present day—the southwestern Germans and the Swiss being markedly broad-headed, while the Scandinavians are as predominantly long-headed.

[1] See Dr. Thurnam "On the Two principal Forms of Ancient British and Gaulish Skulls."

What the natives of Ireland were like at the time of the Roman conquest of Britain, and for centuries afterwards, we have no certain knowledge; but the earliest trustworthy records prove the existence, side by side with one another, of a fair and a dark stock, in Ireland as in Britain. The long form of skull is predominant among the ancient, as among modern, Irish.

II. *The people termed Gauls, and those called Germans, by the Romans, did not differ in any important physical character.*

The terms in which the ancient writers describe both Gauls and Germans are identical. They are always tall people, with massive limbs, fair skins, fierce blue eyes, and hair the colour of which ranges from red to yellow. Zeuss, the great authority on these matters, affirms broadly that no distinction in bodily feature is to be found between the Gauls, the Germans, and the Wends, so far as their characters are recorded by the old historians; and he proves his case by citations from a cloud of witnesses.

An attempt has been made to show that the colour of the hair of the Gauls must have differed very much from that which obtained among the Germans, on the strength of the story told by Suetonius (*Caligula*, 4), that Caligula tried to pass off Gauls for Germans by picking out the tallest, and making then "rutilare et summittere comam."

The Baron de Belloguet remarks upon this passage:

"It was in the very north of Gaul, and near the sea, that Caligula got up this military comedy. And the fact proves that the Belgæ were already sensibly different from their ancestors, whom Strabo had found almost identical with their *brothers* on the other side of the Rhine."

But the fact recorded by Suetonius, if fact it be, proves nothing; for the Germans themselves were in the habit of reddening their hair. Ammianus Marcellinus[1] tells how, in the year 367 A.D., the Roman commander, Jovinus, surprised a body of Alemanni near the town now called Charpeigne, in the valley of the Moselle; and how the Roman soldiers, as, concealed by the thick wood, they stole upon their unsuspecting enemies, saw that some were bathing and others "comas rutilantes ex more." More than two centuries earlier Pliny gives indirect evidence to the same effect when he says of soap:—

"Galliarum hoc inventum rutilandis capillis . . . apud Germanos majore in usu viris quam fœminis."[2]

Here we have a writer who flourished not very long after the date of the Caligula story, telling us that the Gauls invented soap for the purpose of doing that which, according to Suetonius, Caligula forced them to do. And, further

---

[1] *Res Gestæ* xxvii. [2] *Historia Naturalis*, xxviii. 51.

the combined and independent testimony of Pliny and Ammianus assures us that the Germans were as much in the habit of reddening their hair as the Gauls. As to De Belloguet's supposition that, even in Caligula's time, the Gauls had become darker than their ancestors were, it is directly contradicted by Ammianus Marcellinus, who knew the Gauls well. "Celsioris staturæ et candidi pœne Galli sunt omnes, et rutili, luminumque torvitate terribiles," is his description; and it would fit the Gauls who sacked Rome.

III. *In none of the invasions of Britain which have taken place since the Roman dominion, has any other type of man been introduced than one or other of the two which existed during that dominion.*

The North Germans, who effected what is commonly called the Saxon conquest of Britain, were, most assuredly, a fair, yellow, or red-haired, blue-eyed, long-skulled people. So were the Danes and the Norsemen who followed them; though it is very possible that the active slave trade which went on, and the intercourse with Ireland, may have introduced a certain admixture of the dark stock into both Denmark and Norway. The Norman conquest brought in new ethnological elements, the precise value of which cannot be estimated with exactness; but as to their quality, there can be no question, inasmuch as even the wide area from which William drew his followers could yield him nothing but the fair and the dark types of men, already present in Britain. But whether the Norman settlers, on the whole, strengthened the fair or the dark element, is a problem, the elements of the solution of which are not attainable.

I am unable to discover any grounds for believing that a Lapp element has ever entered into the population of these islands. So far as the physical evidence goes, it is perfectly consistent with the hypothesis that the only constituent stocks of that population, now, or at any other period about which we have evidence, are the dark whites, whom I have proposed to call "*Melanochroi*," and the fair whites, or "*Xanthochroi.*"

IV. *The Xanthochroi and the Melanochroi of Britain are, speaking broadly, distributed, at present, as they were in the time of Tacitus; and their representatives on the continent of Europe have the same general distribution as at the earliest period of which we have any record.*

At the present day, and notwithstanding the extensive intermixture effected by the movements consequent on civilization and on political changes, there is a predominance of dark men in the west, and of fair men in the east and north, of Britain. At the present day, as from the earliest times, the predominant constituents of the riverain population of the North Sea and the eastern half of the British Channel, are fair men. The fair stock continues in force through Central Europe, until it is lost in Central Asia. Offshoots of this stock extend into Spain, Italy, and Northern India, and by way of Syria and North Africa, to the Canary Islands. They were known in very early times to the Chinese, and in still earlier to the ancient Egyptians, as frontier tribes. The Thracians were notorious for their fair hair and blue eyes many centuries before our era.

On the other hand, the dark stock predominates in Southern and Western France, in Spain, along the Ligurian shore, and in Western and Southern Italy; in Greece, Asia, Syria, and North Africa; in Arabia, Persia, Afghanistan, and Hindostan, shading gradually, through all stages of darkening, into the type of the modern Egyptian, or of the wild Hill-man of the Dekkan. Nor is there any record of the existence of a different population in all these countries.

The extreme north of Europe, and the northern part of Western Asia, are at present occupied by a Mongoloid stock, and, in the absence of evidence to the contrary, may be assumed to have been so peopled from a very remote epoch. But, as I have said, I can find no evidence that this stock ever took part in peopling Britain. Of the three great stocks of mankind which extend from the western coast of the great Eurasiatic continent to its southern and eastern shores, the Mongoloids occupy a vast triangle, the base of which is the whole of Eastern Asia, while its apex lies in

Lapland. The Melanochroi, on the other hand, may be represented as a broad band stretching from Ireland to Hindostan; while the Xanthochroic area lies between the two, thins out, so to speak, at either end, and mingles, at its margins, with both its neighbours.

Such is a brief and summary statement of what I believe to be the chief facts relating to the physical ethnology of the people of Britain. The conclusions which I draw from these and other facts are—(1) That the Melanochroi and the Xanthochroi are two separate races in the biological sense of the word race; (2) That they have had the same general distribution as at present from the earliest times of which any record exists on the continent of Europe; (3) That the population of the British Islands is derived from them, and from them only.

The people of Europe, however, owe their national names, not to their physical characteristics, but to their languages, or to their political relations; which, it is plain, need not have the slightest relation to these characteristics.

Thus, it is quite certain that, in Cæsar's time, Gaul was divided politically into three nationalities—the Belgæ, the Celtæ, and the Aquitani; and that the last were very widely different, both in language and in physical characteristics, from the two former. The Belgæ and the Celtæ, on the other hand, differed comparatively little either in physique or in language. On the former point there is the distinct testimony of Strabo; as to the latter, St. Jerome states that the "Galatians had almost the same language as the Treviri." Now, the Galatians were emigrant Volcæ Tectosages, and therefore Celtæ; while the Treviri were Belgæ.[1]

At the present day, the physical characters of the people of Belgic Gaul remain distinct from those of the people of Aquitaine, notwithstanding the immense changes which have taken place since Cæsar's time; but Belgæ, Celtæ, and Aquitani (all but a mere fraction of the last two, represented by the Basques and the Bretons) are fused into one nationality, "le peuple Français." But they have adopted the language of one set of invaders, and the name of another; their original names and languages having almost disappeared. Suppose that the French language remained as the sole evidence of the existence of the population of Gaul, would the keenest philologer arrive at any other conclusion than that this population was essentially and fundamentally a " Latin " race, which had had some communication with Celts and Teutons ? Would he so much as suspect the former existence of the Aquitani ?

Community of language testifies to close contact between the people who speak the language, but to nothing else; philology has absolutely nothing to do with ethnology, except so far as it suggests

[1 This proposition is disputed.—1894.]

the existence or the absence of such contact. The contrary assumption, that language is a test of race, has introduced the utmost confusion into ethnological speculation, and has nowhere worked greater scientific and practical mischief than in the ethnology of the British Islands.

What is known, for certain, about the languages spoken in these islands and their affinities may, I believe, be summed up as follows:—

I. *At the time of the Roman conquest, one language, the Celtic, under two principal dialectical divisions, the Cymric and the Gaelic, was spoken throughout the British Islands. Cymric was spoken in Britain, Gaelic*[1] *in Ireland.*

If a language allied to Basque had in earlier times been spoken in the British Islands, there is no evidence that any Euskarian-speaking people remained at the time of the Roman conquest. The dark and the fair population of Britain alike spoke Celtic tongues, and therefore the name "Celt" is as applicable to the one as to the other.

What was spoken in Ireland can only be surmised by reasoning from the knowledge of later times; but there seems to be no doubt that it was Gaelic.

[1 I have been told that the terms "Cymric" and "Gaelic" are antiquated and improper. The reader will please substitute Celtic dialect A and Celtic dialect B for them, and consult, on this subject, especially with regard to proposition III., Professor Rhys' *Early Britain*.—1894.]

II. *The Belgæ and the Celtæ, with the offshoots of the latter in Asia Minor, spoke dialects of the Cymric division of Celtic.*

The evidence of this proposition lies in the statement of St. Jerome before cited; in the similarity of the names of places in Belgic Gaul and in Britain; and in the direct comparison of sundry ancient Gaulish and Belgic words which have been preserved, with the existing Cymric dialects, for which I must refer to the learned work of Brandes.

Formerly, as at the present day, the Cymric dialects of Celtic were spoken by both the fair and the dark stocks.

III. *There is no record of Gaelic being spoken anywhere save in Ireland, Scotland, and the Isle of Man.*

This appears to be the final result of the long discussions which have taken place on this much-debated question. As is the case with the Cymric dialects, Gaelic is now spoken by both dark and fair stocks.

IV. *When the Teutonic languages first became known, they were spoken only*[1] *by Xanthochroi, that is to say, by the Germans, the Scandinavians, and Goths. And they were imported by Xanthochroi into Gaul and into Britain.*

In Gaul, the imported Teutonic dialect has been

[1 "Only" is too strong a word, as there were doubtless some Melanochroi among the Teutonic tribes.—1894.]

completely overpowered by the more or less modified Latin, which it found already in possession; and what Teutonic blood there may be in modern Frenchmen is not adequately represented in their language. In Britain, on the contrary, the Teutonic dialects have overpowered the preexisting forms of speech, and the people are vastly less "Teutonic" than their language. Whatever may have been the extent to which the Celtic-speaking population of the eastern half of Britain was trodden out and supplanted by the Teutonic-speaking Saxons and Danes, it is quite certain that no considerable displacement of the Celtic-speaking people occurred in Cornwall, Wales, or the Highlands of Scotland; and that nothing approaching to the extinction of that people took place in Devonshire, Somerset, or the western moiety of Britain generally. Nevertheless, the fundamentally Teutonic English language is now spoken throughout Britain, except by an insignificant fraction of the population in Wales and the Western Highlands. But it is obvious that this fact affords not the slightest justification for the common practice of speaking of the present inhabitants of Britain as an "Anglo-Saxon" race. It is, in fact, just as absurd as the habit of talking of the French people as a "Latin" race, because they speak a language which is, in the main, derived from Latin. And the absurdity becomes the more patent when those who have no hesitation in calling a Devonshire man, or a Cornish man, an "Anglo-Saxon," would think it ridiculous to call a Tipperary man by the same title, though he and his forefathers may have spoken English for as long a time as the Cornish man.

Ireland, at the earliest period of which we have any knowledge, contained, like Britain, a dark and a fair stock, which, there is every reason to believe, were identical with the dark and the fair stocks of Britain. When the Irish first became known they spoke a Gaelic dialect, and though, for many centuries, Scandinavians made continual incursions upon, and settlements among them, the Teutonic languages made no more way among the Irish than they did among the French. How much Scandinavian blood was introduced there is no evidence to show. But after the conquest of Ireland by Henry II., the English people, consisting in part of the descendants of Cymric speakers, and in part of the descendants of Teutonic speakers, made good their footing in the eastern half of the island, as the Saxons and Danes made good theirs in England; and did their best to complete the parallel by attempting the extirpation of the Gaelic-speaking Irish. And they succeeded to a considerable extent; a large part of Eastern Ireland is now peopled by men who are substantially English by descent, and the English language has spread over the land far beyond the limits of English blood.

Ethnologically, the Irish people were originally, like the people of Britain, a mixture of Melanochroi and Xanthochroi. They resembled the Britons in speaking a Celtic tongue; but it was a Gaelic and not a Cymric form of the Celtic language. Ireland was untouched by the Roman conquest, nor do the Saxons seem to have had any influence upon her destinies, but the Danes and Norsemen poured in a contingent of Teutonism, which has been largely supplemented by English and Scotch efforts.

What, then, is the value of the ethnological difference between the Englishman of the western half of England and the Irishman of the eastern half of Ireland? For what reason does the one deserve the name of a " Celt," and not the other? And further, if we turn to the inhabitants of the western half of Ireland, why should the term "Celts" be applied to them more than to the inhabitants of Cornwall? And if the name is applicable to the one as justly as to the other, why should not intelligence, perseverance, thrift, industry, sobriety, respect for law, be admitted to be Celtic virtues? And why should we not seek for the cause of their absence in something else than the idle pretext of "Celtic blood"?

I have been unable to meet with any answers to these questions.

V. *The Celtic and the Teutonic dialects are members of the same great Aryan family of lan-*

guages; but there is evidence to show that a non-Aryan language was at one time spoken over a large extent of the area occupied by Melanochroi in Europe.

The non-Aryan language here referred to is the Euskarian, now spoken only by the Basques, but which seems in earlier times to have been the language of the Aquitanians and Spaniards, and may possibly have extended much further to the East. Whether it has any connection with the Ligurian and Oscan dialects are questions upon which, of course, I do not presume to offer any opinion. But it is important to remark that it is a language the area of which has gradually diminished without any corresponding extirpation of the people who primitively spoke it; so that the people of Spain and of Aquitaine at the present day must be largely "Euskarian" by descent in just the same sense as the Cornish men are "Celtic" by descent.

Such seem to me to be the main facts respecting the ethnology of the British islands and of Western Europe, which may be said to be fairly established. The hypothesis by which I think (with De Belloguet and Thurnam) the facts may best be explained is this: In very remote times Western Europe and the British islands were inhabited by the dark stock, or the Melanochroi, alone, and these Melanochroi spoke dialects allied to the Euskarian. The Xanthochroi, spreading over the great Eurasiatic plains westward, and speaking Aryan dialects, gradually invaded the territories of the Melanochroi. The Xanthochroi, who thus came into contact with the Western Melanochroi, spoke a Celtic language; and that Celtic language, whether Cymric or Gaelic, spread over the Melanochroi far beyond the limits of intermixture of blood, supplanting Euskarian, just as English and French have supplanted Celtic. Even as early as Cæsar's time, I suppose that the Euskarian was everywhere, except in Spain and in Aquitaine, replaced by Celtic, and thus the Celtic speakers were no longer of one ethnological stock, but of two. Both in Western Europe and in England a third wave of language—in the one case Latin, in the other Teutonic—has spread over the same area. In Western Europe, it has left a fragment of the primary Euskarian in one corner of the country, and a fragment of the secondary Celtic in another. In the British islands, only outlying pools of the secondary linguistic wave remain in Wales, the Highlands, Ireland, and the Isle of Man. If this hypothesis is a sound one, it follows that the name of Celtic is not properly applicable to the Melanochroic or dark stock of Europe. They are merely, so to speak, secondary Celts. The primary and aboriginal Celtic-speaking people are Xanthochroi—the typical Gauls of the ancient writers, and the close allies by blood, customs, and language, of the Germans.

VI

THE ARYAN QUESTION AND PRE-HISTORIC MAN

[1890]

THE rapid increase of natural knowledge, which is the chief characteristic of our age, is effected in various ways. The main army of science moves to the conquest of new worlds slowly and surely, nor ever cedes an inch of the territory gained. But the advance is covered and facilitated by the ceaseless activity of clouds of light troops provided with a weapon—always efficient, if not always an arm of precision—the scientific imagination. It is the business of these *enfants perdus* of science to make raids into the realm of ignorance wherever they see, or think they see, a chance; and cheerfully to accept defeat, or it may be annihilation, as the reward of error. Unfortunately, the public, which watches the progress of the campaign, too often mistakes a dashing incursion of the Uhlans for a forward movement of the main body; fondly imagining that the strategic movement to the rear, which occasionally follows, indicates a battle lost by science. And it must be confessed that the error is too often justified by the effects of the irrepressible tendency which men of science share with all other sorts of men known to me, to be impatient of that most wholesome state of mind—suspended judgment; to assume the objective truth of speculations which, from the nature of the evidence in their favour, can have no claim to be more than working hypotheses.

The history of the "Aryan question" affords a striking illustration of these general remarks.

About a century ago, Sir William Jones pointed out the close alliance of the chief European languages with Sanskrit and its derivative dialects now spoken in India. Brilliant and laborious philologists, in long succession, enlarged and strengthened this position, until the truth that Sanskrit, Zend, Armenian, Greek, Latin, Lithuanian, Slavonian, German, Celtic, and so on, stand to one another in the relation of descendants from a common stock, became firmly established, and thenceforward formed part of the permanent acquisitions of science. Moreover, the term "Aryan" is very generally, if not universally, accepted as a name for the group of languages thus allied. Hence, when one speaks of "Aryan languages," no hypothetical assumptions are involved. It is a matter of fact that such languages exist, that they present certain substantial and formal relations, and that convention sanctions the name applied to them. But the close connection of these widely differentiated languages remains altogether inexplicable, unless it is admitted that they are modifications of an original relatively undifferentiated tongue; just as the intimate affinities of the Romance languages—French, Italian, Spanish, and the rest—would be incomprehensible if there were no Latin. The original or "primitive Aryan" tongue, thus postulated, unfortunately no longer exists. It is a hypothetical entity, which corresponds with the "primitive stock" of generic and higher groups among plants and animals; and the acknowledgment of its former existence, and of the process of evolution which has brought about the present state of things philological, is forced upon us by deductive reasoning of similar cogency to that employed about things biological.

Thus, the former existence of a body of relatively uniform dialects, which may be called primitive Aryan, may be added to the stock of definitely acquired truths. But it is obvious that, in the absence of writing or of phonographs, the existence of a language implies that of speakers. If there were primitive Aryan dialects, there must have been primitive Aryan people who used them; and these people must have resided somewhere or other on the earth's surface. Hence philology, without stepping beyond its legitimate bounds and keeping speculation within the limits of bare necessity, arrives, not only at the conceptions of Aryan languages and of a primitive Aryan language; but of a primitive Aryan people and of a primitive Aryan home, or country occupied by them.

But where was this home of the Aryans? When the labours of modern philologists began, Sanskrit was the most archaic of all the Aryan languages known to them. It appeared to present the qualifications required in the parental or primitive Aryan. Brilliant Uhlans made a charge at this opening. The scientific imagination seated the primitive Aryans in the valley of the Ganges; and showed, as in a vision the successive columns, guided by enterprising Brahmins, which set out thence to people the regions of the western world with Greeks and Celts and Germans. But the progress of philology itself sufficed to show that this Balaclava charge, however magnificent, was not profitable warfare. The internal evidence of the Vedas proved that their composers had not reached the Ganges. On the other hand, the comparison of Zend with Sanskrit left no alternative open to the assumption that these languages were modifications of an original Indo-Iranian tongue, spoken by a people of whom the Aryans of India and those of Persia were offshoots, and who could therefore be hardly lodged elsewhere than on the frontiers of both Persia and India—that is to say, somewhere in the region which is at present known under the names of Turkestan, Afghanistan, and Kafiristan. Thus far, it can hardly be doubted that we are well within the ground of which science has taken enduring possession. But the Uhlans were not content to remain within the lines of this surely-won position. For some reason, which is not quite clear to me, they thought fit to restrict the home of the primitive Aryans to a particular part of the region in question; to lodge them amidst the bleak heights of the long range of the Hindoo Koosh and on the inhospitable plateau of Pamir. From their hives in these secluded valleys and wind-swept wastes, successive swarms of Celts and Greco-Latins, Teutons and Slavs, were thrown off to settle, after long wanderings, in distant Europe. The Hindoo-Koosh-Pamir theory, once enunciated, gradually hardened into a sort of dogma; and there have not been wanting theorists, who laid down the routes of the successive bands of emigrants with as much confidence as if they had access to the records of the office of a primitive Aryan Quartermaster-General. It is really singular to observe the deference which has been shown, and is yet sometimes shown, to a speculation which can, at best, claim to be regarded as nothing better than a somewhat risky working hypothesis.

Forty years ago, the credit of the Hindoo-Koosh-Pamir theory had risen almost to that of an axiom. The first person to instil doubt of its value into my mind was the late Robert Gordon Latham, a man of great learning and singular originality, whose attacks upon the Hindoo-Kooshite doctrine could scarcely have failed as completely as they did, if his great powers had been bestowed upon making his books not only worthy of being read, but readable. The impression left upon my mind, at that time, by various conversations about the "Sarmatian hypothesis," which my friend wished to substitute for the Hindoo-Koosh-Pamir speculation, was that the one and the other rested pretty much upon a like foundation of guess-work. That there was no sufficient reason for planting the primitive Aryans in the Hindoo Koosh, or in Pamir, seemed plain enough; but that there was little better ground, on the evidence then adduced, for settling them in the region at present occupied by Western Russia, or Podolia, appeared to me to be not less plain. The most I thought Latham proved was, that the Aryan people of Indo-Iranian speech were just as likely to have come from Europe, as the Aryan people of Greek, or Teutonic, or Celtic speech from Asia. Of late years, Latham's views, so long neglected, or mentioned merely as an example of insular eccentricity, have been taken up and advocated with much ability in Germany as well as in this country—principally by philologists. Indeed, the glory of Hindou-Koosh-Pamir seems altogether to have departed. Professor Max Müller, to whom Aryan philology owes so much, will not say more now, than that he holds by the conviction that the seat of the primitive Aryans was "somewhere in Asia." Dr. Schrader sums up in favour of European Russia; while Herr Penka would have us transplant the home of the primitive Aryans from Pamir in the far east to the Scandinavian peninsula in the far west.

I must refer those who desire to acquaint themselves with the philological arguments on which these conclusions are based to the recently published works of Dr. Schrader and Canon Taylor;[1] and to Penka's "Die Herkunft der Arier," which, in spite of the strong spice of the Uhlan which runs through it, I have found extremely well worth study. I do not pretend to be able to look at the Aryan question under any but the biological aspect; to which I now turn.

Any biologist who studies the history of the Aryan question, and, taking the philological facts on trust, regards it exclusively from the point of view of anthropology, will observe that, very early, the purely biological conception of "race"

[1] Schrader, *Prehistoric Antiquities of the Aryan Peoples*. Translated by F. B. Jevons, M.A., 1890. Taylor, *The Origin of the Aryans*, 1890.

illegitimately mixed itself up with the ideas derived from pure philology. It is quite proper to speak of Aryan "people," because, as we have seen, the existence of the language implies that of a people who speak it; it might be equally permissible to call Latin people all those who speak Romance dialects. But, just as the application of the term Latin "race" to the divers people who speak Romance languages, at the present day, is none the less absurd because it is common; so, it is quite possible, that it may be equally wrong to call the people who spoke the primitive Aryan dialects and inhabited the primitive home, the Aryan race. "Aryan" is properly a term of classification used in philology. "Race" is the name of a sub-division of one of those groups of living things which are called "species" in the technical language of Zoology and Botany; and the term connotes the possession of characters distinct from those of the other members of the species, which have a strong tendency to appear in the progeny of all members of the races. Such race-characters may be either bodily or mental, though in practice, the latter, as less easy of observation and definition, can rarely be taken into account. Language is rooted half in the bodily and half in the mental nature of man. The vocal sounds which form the raw materials of language could not be produced without a peculiar conformation of the organs of speech; the enunciation of duly accented syllables would be impossible without the nicest co-ordination of the action of the muscles which move these organs; and such co-ordination depends on the mechanism of certain portions of the nervous system. It is therefore conceivable that the structure of this highly complex speaking apparatus should determine a man's linguistic potentiality; that is to say, should enable him to use a language of one class and not of another. It is further conceivable that a particular linguistic potentiality should be inherited and become as good a race mark as any other. As a matter of fact, it is not proven that the linguistic potentialities of all men are the same. It is affirmed, for example, that, in the United States, the enunciation and the timbre of the voice of an American-born negro, however thoroughly he may have learned English, can be readily distinguished from that of a white man. But, even admitting that differences may obtain among the various races of men, to this extent, I do not think that there is any good ground for the supposition that an infant of any race would be unable to learn, and to use with ease, the language of any other race of men among whom it might be brought up. History abundantly proves the transmission of languages from some races to others; and there is no evidence, that I know of, to show that any race is incapable of substituting a foreign idiom for its native tongue.

From these considerations it follows that community of language is no proof of unity of race, is not even presumptive evidence of racial identity.[1] All that it does prove is that, at some time or other, free and prolonged intercourse has taken place between the speakers of the same language. Philology, therefore, while it may have a perfect right to postulate the existence of a primitive Aryan "people," has no business to substitute "race" for "people." The speakers of primitive Aryan may have been a mixture of two or more races, just as are the speakers of English and of French, at the present time.

The older philological ethnologists felt the difficulty which arose out of their identification of linguistic with racial affinity, but were not dismayed by it. Strong in the prestige of their great discovery of the unity of the Aryan tongues, they were quite prepared to make the philological and the biological categories fit, by the exercise of a little pressure on that about which they knew less. And their judgment was often un-

[1] Canon Taylor (*Origin of the Aryans*, p 31) states that "Cuno .... was the first to insist on what is now looked on as an axiom in ethnology—that race is not co-extensive with language," in a work published in 1871. I may be permitted to quote a passage from a lecture delivered on the 9th of January, 1870, which brought me into a great deal of trouble. "Physical, mental, and moral peculiarities go with blood and not with language. In the United States the negroes have spoken English for generations; but no one on that ground would call them Englishmen, or expect them to differ physically, mentally, or morally from other negroes."—*Pall Mall Gazette*, Jan. 10, 1870. But the "axiom in ethnology" had been implied, if not enunciated, before my time; for example, by Desmoulins in 1826 (See above p. 215).

consciously warped by strong monogenistic proclivities, which, at bottom, however respectable and philanthropic their origin, had nothing to do with science. So the patent fact that men of Aryan speech presented widely diverse racial characters was explained away by maintaining that the physical differentiation was post-Aryan; to put it broadly, that the Aryans in Hindoo-Koosh-Pamir were truly of one race; but that, while one colony, subjected to the sweltering heat of the Gangetic plains, had fined down and darkened into the Bengalee, another had bleached and shot up, under the cool and misty skies of the north, into the semblance of Pomeranian Grenadiers; or of blue-eyed, fair-skinned, six-foot Scotch Highlanders. I do not know that any of the Uhlans who fought so vigorously under this flag are left now. I doubt if any one is prepared to say that he believes that the influence of external conditions, alone, accounts for the wide physical differences between Englishmen and Bengalese. So far as India is concerned, the internal evidence of the old literature sufficiently proves that the Aryan invaders were "white" men. It is hardly to be doubted that they intermixed with the dark Dravidian aborigines; and that the high-caste Hindoos are what they are in virtue of the Aryan blood which they have inherited,[1] and of

[1] I am unable to discover good grounds for the severity of the criticism, in the name of "the anthropologists," with which Professor Max Müller's assertion that the same blood runs in the

the selective influence of their surroundings operating on the mixture.

The assumption that, as there must have been a primitive Aryan people, in the philological sense, so that people must have constituted a race in the biological sense, is pretty generally made in modern discussions of the Aryan problem. But whether the men of the primitive Aryan race were blonds or brunets, whether they had long or round heads, were tall or were short, are hotly debated questions, into the discussion of which considerations quite foreign to science are sometimes imported. The combination of swarthiness with stature above the average and a long skull, confer upon me the serene impartiality of a mongrel; and, having given this pledge of fair dealing, I proceed to state the case for the hypothesis I am inclined to adopt. In doing so, I am aware that I deliberately take the shilling of the recruiting sergeant of the Light Brigade, and I warn all and sundry that such is the case.

Looking at the discussions which have taken

veins of English soldiers "as in the veins of the dark Bengalese," and that there is "a legitimate relationship between Hindoo, Greek, and Teuton," has been visited. So far as I know anything about anthropology, I should say that these statements may be correct literally, and probably are so substantially. I do not know of any good reason for the physical differences between a high-caste Hindoo and a Dravidian, except the Aryan blood in the veins of the former; and the strength of the infusion is probably quite as great in some Hindoos as in some English soldiers.

place from a purely anthropological point of view, the first point which has struck me is that the problem is far more complicated and difficult than many of the disputants appear to imagine; and the second, that the data upon which we have to go are grievously insufficient in extent and in precision. Our historical records cover such an infinitesimally small extent of the past life of humanity, that we obtain little help from them. Even so late as 1500 B.C., northern Eurasia lies in historical darkness, except for such glimmer of light as may be thrown here and there by the literatures of Egypt and of Babylonia. Yet, at that time, it is probable that Sanskrit, Zend, and Greek, to say nothing of other Aryan tongues, had long been differentiated from primitive Aryan. Even a thousand years later, little enough accurate information is to be had about the racial characters of the European and Asiatic tribes known to the Greeks. We are thrown upon such resources as archæology and human palæontology have to offer, and notwithstanding the remarkable progress made of late years, they are still meagre. Nevertheless, it strikes me that, from the purely anthropological side, there is a good deal to be said in favour of the two propositions maintained by the new school of philologists; first, that the people who spoke "primitive Aryan" were a distinct and well-marked race of mankind; and, secondly, that the area of the distribution of this race, in primæval times, lay in Europe, rather than in Asia.

For the last two thousand years, at least, the southern half of Scandinavia and the opposite or southern shores of the Baltic have been occupied by a race of mankind possessed of very definite characters. Typical specimens have tall and massive frames, fair complexions, blue eyes, and yellow or reddish hair—that is to say, they are pronounced blonds. Their skulls are long, in the sense that the breadth is usually less, often much less, than four-fifths of the length, and they are usually tolerably high. But in this last respect they vary. Men of this blond, long-headed race abound from eastern Prussia to northern Belgium; they are met with in northern France and are common in some parts of our own islands. The people of Teutonic speech, Goths, Saxons, Alemanni, and Franks, who poured forth out of the regions bordering the North Sea and the Baltic, to the destruction of the Roman Empire, were men of this race; and the accounts of the ancient historians of the incursions of the Gauls into Italy and Greece, between the fifth and the second centuries B.C., leave little doubt that their hordes were largely, if not wholly, composed of similar men. The contents of numerous interments in southern Scandinavia prove that, as far back as archæology takes us into the so-called neolithic age, the great majority of the inhabitants had the same stature and cranial peculiarities as at present, though their bony fabric bears marks of somewhat greater ruggedness and savagery. There is no evidence that the country was occupied by men before the advent of these tall, blond long-heads. But there is proof of the presence, along with the latter, of a small percentage of people with broad skulls; skulls, that is, the breadth of which is more, often very much more, than four-fifths of the length.

At the present day, in whatever direction we travel inland from the continental area occupied by the blond long-heads, whether south-west, into central France; south, through the Walloon provinces of Belgium into eastern France; into Switzerland, South Germany, and the Tyrol; or south-east, into Poland and Russia; or north, into Finland and Lapland, broad-heads make their appearance, in force, among the long-heads. And, eventually, we find ourselves among people who are as regularly broad-headed as the Swedes and North Germans are long-headed. As a general rule, in France, Belgium, Switzerland, and South Germany, the increase in the proportion of broad skulls is accompanied by the appearance of a larger and larger proportion of men of brunet complexion and of a lower stature; until, in central France and thence eastwards, through the Cevennes and the Alps of Dauphiny, Savoy, and Piedmont, to the western plains of North Italy, the *tall blond long-heads*[1] practically disappear, and are replaced by *short brunet broad-heads*. The ordinary Savoyard may be described in terms the converse of those which apply to the ordinary Swede. He is short, swarthy, dark-eyed, dark-haired, and his skull is very broad. Between the two extreme types, the one seated on the shores of the North Sea and the Baltic, and the other on those of the Mediterranean, there are all sorts of intermediate forms, in which breadth of skull may be found in tall and in short blond men, and in tall brunet men.

There is much reason to believe that the brunet broad-heads, now met with in central France and in the west central European highlands, have inhabited the same region, not only throughout the historical period, but long before it commenced; and it is probable that their area of occupation was formerly more extensive. For, if we leave

---

[1] I may plead the precedent of the good English words "block-head" and "thick-head" for "broad-head" and "long-head," but I cannot say that they are elegant. · I might have employed the technical terms brachycephali and dolichocephali. But it cannot be said that they are much more graceful; and, moreover, they are sometimes employed in senses different from that which I have given in the definition of broad-heads and long heads. The *cephalic index* is a number which expresses the relation of the breadth to the length of a skull, taking the latter as 100. Therefore "broad-heads" have the cephalic index above 80 and "long-heads" have it below 80. The physiological value of the difference is unknown; its morphological value depends upon the observed fact of the constancy of the occurrence of either long skulls or broad skulls among large bodies of mankind.

aside the comparatively late incursions of the Asiatic races, the centre of eruption of the invaders of the southern moiety of Europe has been situated in the north and west. In the case of the Teutonic inroads upon the Empire of Rome, it undoubtedly lay in the area now occupied by the blond long-heads; and, in that of the antecedent Gaulish invasions, the physical characters ascribed to the leading tribes point to the same conclusion. Whatever the causes which led to the breaking out of bounds of the blond long-heads, in mass, at particular epochs, the natural increase in numbers of a vigorous and fertile race must always have impelled them to press upon their neighbours, and thereby afford abundant occasions for intermixture. If, at any given pre-historic time, we suppose the lowlands verging on the Baltic and the North Sea to have been inhabited by pure blond long-heads, while the central highlands were occupied by pure brunet short-heads, the two would certainly meet and intermix in course of time, in spite of the vast belt of dense forest which extended, almost uninterruptedly, from the Carpathians to the Ardennes; and the result would be such an irregular gradation of the one type into the other as we do, in fact, meet with.

On the south-east, east, and north-east, throughout what was once the kingdom of Poland, and in Finland, the preponderance of broad heads goes along with a wide prevalence of blond complexion and of good stature. In the extreme north, on the other hand, marked broad-headedness is combined with low stature, swarthiness, and more or less strongly mongolian features, in the Lapps. And it is to be observed that this type prevails increasingly to the eastward, among the central Asiatic populations.

The population of the British Islands, at the present time, offers the two extremes of the tall blond and the short brunet types. The tall blond long-heads resemble those of the continent; but our short brunet race is long-headed. Brunet broad-heads, such as those met with in the central European highlands, do not exist among us. This absence of any considerable number of distinctly broad-headed people (say with the cephalic index above 81 or 82) in the modern population of the United Kingdom is the more remarkable, since the investigations of the late Dr. Thurnam, and others, proved the existence of a large proportion of tall broad-heads among the people interred in British tumuli of the neolithic age. It would seem that these broad-skulled immigrants have been absorbed by an older long-skulled population; just as, in South Germany, the long-headed Alemanni have been absorbed by the older broad-heads. The short brunet long-heads are not peculiar to our islands. On the contrary, they abound in western France and in Spain, while they predominate in Sardinia, Corsica, and South Italy, and, it may be, occupied a much larger area in ancient times.

Thus, in the region which has been under consideration, there are evidences of the existence of four races of men—(1) blond long-heads of tall stature, (2) brunet broad-heads of short stature, (3) mongoloid brunet broad-heads of short stature, (4) brunet long-heads of short stature. The regions in which these races appear with least admixture are—(1) Scandinavia, North Germany, and parts of the British Islands; (2) central France, the central European highlands, and Piedmont; (3) Arctic and eastern Europe, central Asia; (4) the western parts of the British Islands and of France; Spain, South Italy. And the inhabitants of the localities which lie between these foci present the intermediate gradations, such as short blond long-heads, and tall brunet short-heads and long-heads which might be expected to result from their intermixture. The evidence at present extant is consistent with the supposition that the blond long-heads, the brunet broad-heads, and the brunet long-heads have existed in Europe throughout historic times, and very far back into pre-historic times. There is no proof of any migration of Asiatics into Europe, west of the basin of the Dnieper, down to the time of Attila. On the contrary, the first great movements of the European population of which there is any conclusive evidence is that series of Gaulish invasions of the east and south, which ultimately extended from North Italy as far as Galatia in Asia Minor.

It is now time to consider the relations between the phenomena of racial distribution, as thus defined, and those of the distribution of languages. The blond long-heads of Europe speak, or have spoken, Lithuanian, Teutonic, or Celtic dialects, and they are not known to have ever used any but these Aryan languages. A large proportion of the brunet broad-heads once spoke the Ligurian and the Rhætic dialects, which are believed to have been non-Aryan. But, when the Romans made acquaintance with Transalpine Gaul, the inhabitants of that country between the Garonne and the Seine (Cæsar's *Celtica*) seem, at any rate for the most part, to have spoken Celtic dialects. The brunet long-heads of Spain and of France appear to have used a non-Aryan language, that Euskarian which still lives on the shores of the Bay of Biscay. In Britain there is no certain knowledge of their use of any but Celtic tongues. What they spoke in the Mediterranean islands and in South Italy does not appear.

The blond broad-heads of Poland and West Russia form part of a people who, when they first made their appearance in history, occupied the marshy plains imperfectly drained by the Vistula, on the west, the Duna, on the north, and the Dnieper and Bug, on the south. They were known to their neighbours as Wends, and among themselves as Serbs and Slavs. The Slavonic languages spoken by these people are said to be most closely allied to that of the Lithuanians, who lay upon their northern border. The Slavs resemble the South Germans in the predominance of broad-heads among them, while stature and complexion vary from the, often tall, blonds who prevail in Poland and great Russia to the, often short, brunets common elsewhere. There is certainly nothing in the history of the Slav people to interfere with the supposition that, from very early times, they have been a mixed race. For their country lies between that of the tall blond long-heads on the north, that of the short brunet broad-heads of the European type on the west, and that of the short brunet broad-heads of the Asiatic type on the east: and, throughout their history, they have either thrust themselves among their neighbours, or have been overrun and trampled down by them. Gauls and Goths have traversed their country, on their way to the east and south: Finno-tataric people, on their way to the west, have not only done the like, but have held them in subjection for centuries. On the other hand, there have been times when their western frontier advanced beyond the Elbe; indeed, it is asserted that they have sent colonies to Holland and even as far as southern England. A large part of eastern Germany; Bohemia,

Moravia, Hungary; the lower valley of the Danube and the Balkan peninsula, have been largely or completely Slavonised; and the Slavonic rule and language, which once had trouble to hold their own in West Russia and Little Russia, have now extended their sway over all the Finno-tataric populations of Great Russia; while they are advancing, among those of central Asia, up to the frontiers of India on the south and to the Pacific on the extreme east. Thus it is hardly possible that fewer than three races should have contributed to the formation of the Slavonic people; namely, the blond long-heads, the European brunet broad-heads, and the Asiatic brunet broad-heads. And, in the absence of evidence to the contrary, it is certainly permissible to suppose that it is the first race which has furnished the blond complexion and the stature observable in so many, especially of the northern Slavs, and that the brunet complexion and the broad skulls must be attributed to the other two. But, if that supposition is permissible, then the Aryan form and substance of the Slavonic languages may also be fairly supposed to have proceeded from the blond long-heads. They could not have come from the Asiatic brunet broad-heads, who all speak non-Aryan languages; and the presumption is against their coming from the brunet broad-heads of the central European highlands, among whom an apparently non-Aryan language was largely spoken, even in historical times.

In the same way, the tall blond tribes among the Fins may be accounted for as the product of admixture. The great majority of the Finno-tataric people are brunet broad-heads of the Asiatic type. But that the Fins proper have long been in contact with Aryans is evidenced by the many words borrowed from Aryan which their language contains. Hence there has been abundant opportunity for the mixture of races; and for the transference to some of the Fins of more or fewer of the physical characters of the Aryans and *vice versâ*. On any hypothesis, the frontier between Aryan and Finno-tataric people must have extended across west-central Asia for a very long period; and, at any point of this frontier, it has been possible that mixed races of blond Fins or of brunet Aryans should be formed.

So much for the European people who now speak Celtic, or Teutonic, or Slavonian, or Lithuanian tongues; or who are known to have spoken them, before the supersession of so many of the early native dialects by the Romance modifications of the language of Rome. With respect to the original speakers of Greek and Latin, the unravelling of the tangled ethnology of the Balkan peninsula and the ordering of the chaos of that of Italy are enterprises upon which I do not propose to enter. In regard to the first, however, there are a few tolerably satisfactory data. The ancient Thracians were proverbially blue-eyed and fair-haired. Tall blonds were common among the ancient Greeks, who were a long-headed people; and the Sphakiots of Crete, probably the purest representatives of the old Hellenes in existence, are tall and blond. But considering that Greek colonisation was taking place on a great scale in the eighth century B.C., and that, centuries earlier and later, the restless Hellene had been fighting, trading, plundering and kidnapping, on both sides of the Ægean, and perhaps as far as the shores of Syria and of Egypt, it is probable that, even at the dawn of history, the maritime Greeks were a very mixed race. On the other hand, the Dorians may well have preserved the original type; and their famous migration may be the earliest known example of those movements of the Aryan race which were, in later times, to change the face of Europe. Analogy perhaps justifies a guess, that those ethnological shadows, the Pelasgi, may have been an earlier mixed population, like that of Western Gaul and of Britain before the Teutonic invasion. At any rate, the tall blond long-heads are so well represented in the oldest history of the Balkan peninsula, that they may be credited with the Aryan languages spoken there. And it may be that the tradition which peopled Phrygia with Thracians represents a real movement of the Aryan race into Asia Minor, such as that which in after years carried the Gauls thither.

The difficulties in the way of a probable identification of the people among whom the various dialects of the Latin group developed themselves, with any race traceable in Italy in historical times, are very great. In addition to the Italic "aborigines" northern Italy was peopled by Ligurian brunet broad-heads; with Gauls, probably, to a large extent, blond long-heads; with Illyrians, about whom nothing is known. Besides these, there were those perplexing people the Etruscans, who seem to have been, originally, brunet long-heads. South Italy and Sicily present a contingent of "Sikels," Phœnicians and Greeks; while over all, in comparatively modern times, follows a wash of Teutonic blood. The Latin dialects arose, no one knows how, among the tribes of Central Italy, encompassed on all sides by people of the most various physical characters, who were gradually absorbed into the eternally widening maw of Rome, and there, by dint of using the same speech, became the first example of that wonderful ethnological hotchpotch miscalled the Latin race. The only trustworthy guide here is archæological investigation. A great advance will have been made when the race characters of the pre-historic people of the terremare (who are identified by

Helbig[1] with the primitive Umbrians) become fully known.

I cannot learn that the ancient literatures of India and of Persia give any definite information about the complexion of the Indo-Iranians, beyond conveying the impression that they were what we vaguely call white men. But it is important to note that tall blond people make their appearance sporadically among the Tadjiks of Persia and of Turkestan; that the Siah-posh and Galtchas of the mountainous barrier between Turkestan and India are such; and that the same characters obtain largely among the Kurds on the western frontier of Persia, at the present day. The Kurds and the Galtchas are generally broad-headed, the others are long-headed. These people and the ancient Alans thus form a series of stepping-stones between the blond Aryans of Europe and those of Asia, standing up amidst the flood of Finno-tataric people which has inundated the rest of the interval between the sources of the Dnieper and those of the Oxus. If only more was known about the Sarmatians and the Scythians of the oldest historians, it is not improbable, I think, that we should discover that, even in historical times, the area occupied by the blond long-heads

---

[1] *Die Italiker in der Poebene*, 1879. See for much valuable information respecting the races of the Balkan and Italic peninsulæ, Zampa's essay, "Vergleichende Anthropologische Ethnographie von Apulien," *Zeitschrift für Ethnologie*, xviii., 1886.

of Aryan speech has been, at least temporarily, continuous from the shores of the North Sea to central Asia.

Suppose it to be admitted, as a fair working hypothesis, that the blond long-heads once extended without a break over this vast area, and that all the Aryan tongues have been developed out of their original speech, the question respecting the home of the race when the various families of Aryan speech were in the condition of inceptive dialects remains open. For all that, at first, appears to the contrary, it may have been in the west, or in the east, or anywhere between the two. In seeking for a solution of this obscure problem, it is an important preliminary to grasp the truth that the Aryan race must be much older than the primitive Aryan speech. It is not to be seriously imagined that the latter sprang suddenly into existence, by the act of a jealous Deity, apparently unaware of the strength of man's native tendency towards confusion of speech. But if all the diverse languages of men were not brought suddenly into existence, in order to frustrate the plans of the audacious bricklayers of the plain of Shinar; if this professedly historical statement is only another "type," and primitive Aryan, like all other languages, was built up by a secular process of development, the blond long-heads, among whom it grew into shape, must for ages have been, philologically speaking, non-Aryans, or perhaps one should say "pro-Aryans." I suppose it may be safely assumed that Sanskrit and Zend and Greek were fully differentiated in the year 1500 B.C. If so, how much further back must the existence of the primitive Aryan, from which these proceeded, be dated? And how much further yet, that real *juventus mundi* (so far as man is concerned) when primitive Aryan was in course of formation? And how much further still, the differentiation of the nascent Aryan blond long-head race from the primitive stock of mankind?

If any one maintains that the blond long-headed people, among whom, by the hypothesis, the primitive Aryan language was generated may have formed a separate race as far back as the pleistocene epoch, when the first unquestionable records of man make their appearance, I do not see that he goes beyond possibility—though, of course, that is a very different thing from proving his case. But, if the blond long-heads are thus ancient, the problem of their primitive seat puts on an altogether new aspect. Speculation must take into account climatal and geographical conditions widely different from those which obtain in northern Eurasia at the present day. During much of the vast length of the pleistocene period, it would seem that men could no more have lived either in Britain north of the Thames, or in

Scandinavia, or in northern Germany, or in northern Russia, than they can live now in the interior of Greenland, seeing that the land was covered by a great ice sheet like that which at present shrouds the latter country. At that epoch, the blond long-heads cannot reasonably be supposed to have occupied the regions in which we meet with them in the oldest times of which history has kept a record.

But even if we are content to assume a vastly less antiquity for the Aryan race; if we only make the assumption, for which there is considerable positive warranty, that it has existed in Europe ever since the end of the pleistocene period—when the fauna and flora assumed approximately their present condition and the state of things called Recent by geologists set in—we have to reckon with a distribution of land and water, not only very different from that which at present obtains in northern Eurasia, but of such a nature that it can hardly fail to have exerted a great influence on the development and the distribution of the races of mankind. (See page 250, note [2].)

At the present time, four great separate bodies of water, the Black Sea, the Caspian, the Sea of Aral, and Lake Balkash, occupy the southern end of the vast plains which extend from the Arctic Sea to the highlands of the Balkan peninsula, of Asia Minor, of Persia, of Afghanistan, and of the high plateaus of central Asia as far as the Altai.

They lie for the most part between the parallels of 40° and 50° N. and are separated by wide stretches of barren and salt-laden wastes. The surface of Balkash is 514 feet, that of the Aral 158 feet above the Mediterranean, that of the Caspian eighty-five feet below it. The Black Sea is in free communication with the Mediterranean by the Bosphorus and the Dardanelles; but the others, in historical times, have been, at most, temporarily connected with it and with one another, by relatively insignificant channels. This state of things, however, is comparatively modern. At no very distant period, the land of Asia Minor was continuous with that of Europe, across the present site of the Bosphorus, forming a barrier several hundred feet high, which dammed up the waters of the Black Sea. A vast extent of eastern Europe and of western central Asia thus became a huge reservoir, the lowest part of the lip of which was probably situated somewhat more than 200 feet above the sea level, along the present southern watershed of the Obi, which flows into the Arctic Ocean. Into this basin, the largest rivers of Europe, such as the Danube and the Volga, and what were then great rivers of Asia, the Oxus and Jaxartes, with all the intermediate affluents, poured their waters. In addition, it received the overflow of Lake Balkash, then much larger; and, probably, that of the inland sea of Mongolia. At that time, the level of the Sea of Aral stood at least 60 feet higher than it does at present.[1] Instead of the separate Black, Caspian, and Aral seas, there was one vast Ponto-Aralian Mediterranean, which must have been prolonged into arms and fiords along the lower valleys of the Danube, the Volga (in the course of which Caspian shells are now found as far as the Kuma), the Ural, and the other affluent rivers—while it seems to have sent its overflow, northward, through the present basin of the Obi. At the same time, there is reason to believe that the northern coast of Asia, which everywhere shows signs of recent slow upheaval, was situated far to the south of its present position. The consequences of this state of things have an extremely important bearing on the question under discussion. In the first place, an insular climate must be substituted for the present extremely continental climate of west central Eurasia. That is an important fact in many ways. For example, the present eastern climatal limitations of the beech could not have existed, and if primitive Aryan goes back thus far, the arguments based upon the occurrence of its name in some Aryan languages and not in others lose their force. In the second place, the European and the Asiatic moieties of the great Eurasiatic

[1] This is proved by the old shore-marks on the hill of Kashkanatao in the midst of the delta of the Oxus. Some authorities put the ancient level very much higher—200 feet or more (Keane, *Asia*, p. 408).

plains were cut off from one another by the Ponto-Aralian Mediterranean and its prolongations. In the third place, direct access to Asia Minor, to the Caucasus, to the Persian highlands, and to Afghanistan, from the European moiety was completely barred; while the tribes of eastern central Asia were equally shut out from Persia and from India by huge mountain ranges and table lands. Thus, if the blond long-head race existed so far back as the epoch in which the Ponto-Aralian Mediterranean had its full extension, space for its development, under the most favourable conditions, and free from any serious intrusion of foreign elements from Asia, was presented in northern and eastern Europe.

When the slow erosion of the passage of the Dardanelles drained the Ponto-Aralian waters into the Mediterranean, they must have everywhere fallen as near the level of the latter as the make of the country permitted, remaining, at first, connected by such straits as that of which the traces yet persist between the Black and the Caspian, the Caspian and the Aral Seas respectively. Then, the gradual elevation of the land of northern Siberia, bringing in its train a continental climate, with its dry air and intense summer heats, the loss by evaporation soon exceeded the greatly reduced supply of water, and Balkash, Aral, and Caspian gradually shrank to their present dimensions. In the course of this process, the broad plains between the separated inland seas, as soon as they were laid bare, threw open easy routes to the Caucasus and to Turkestan, which might well be utilised by the blond long-heads moving eastward through the plains, contemporaneously left dry, south and east of the Ural chain. The same process of desiccation, however, would render the route from east central Asia westward as easily practicable; and, in the end, the Aryan stock might easily be cut in two, as we now find it to be, by the movement of the Mongoloid brunet broad-heads to the west.

Thus we arrive at what is practically Latham's Sarmatian hypothesis—if the term "Sarmatian" is stretched a little, so as to include the higher parts and a good deal of the northern slopes of Europe between the Ural and the German Ocean; an immense area of country, at least as large as that now included between the Black Sea, the Atlantic, the Baltic, and the Mediterranean.

If we imagine the blond long-head race to have been spread over this area, while the primitive Aryan language was in course of formation, its north-western and its south-eastern tribes will have been 1,500, or more, miles apart. Thus, there will have been ample scope for linguistic differentiation; and, as adjacent tribes were probably influenced by the same causes, it is reasonable to suppose that, at any given region of the periphery the process of differentiation, whether brought about by internal or external agencies, will have been analogous. Hence, it is permissible to imagine that, even before primitive Aryan had attained its full development, the course of that development had become somewhat different in different localities; and, in this sense, it may be quite true that one uniform primitive Aryan language never existed. The nascent mode of speech may very early have got a twist, so to speak, towards Lithuanian, Slavonian, Teutonic, or Celtic, in the north and west; towards Thracian and Greek, in the south-west; towards Armenian in the south; towards Indo-Iranian in the south-east. With the centrifugal movements of the several fractions of the race, these tendencies of peripheral groups would naturally become more and more intensified in proportion to their isolation. No doubt, in the centre and in other parts of the periphery of the Aryan region, other dialectic groups made their appearance; but whatever development they may have attained, these have failed to maintain themselves in the battle with the Finno-tataric tribes, or with the stronger among their own kith and kin.[1]

Thus I think that the most plausible hypothetical answers which can be given to the two questions which we put at starting are these.

[1] See the views of J. Schmidt (stated and discussed in Schrader and Jevons, pp. 63-67), with which those here set forth are substantially identical.

There was and is an Aryan race—that is to say, the characteristic modes of speech, termed Aryan, were developed among the blond long-heads alone however much some of them may have been modified by the importation of non-Aryan elements. As to the " home " of the Aryan race, it was in Europe, and lay chiefly east of the central highlands and west of the Ural. From this region it spread west, along the coasts of the North Sea to our islands, where, probably, it met the brunet long-heads; to France, where it found both these and the brunet short-heads; to Switzerland and South Germany, where it impinged on the brunet short-heads; to Italy, where brunet short-heads seem to have abounded in the north and long-heads in the south; and to the Balkan peninsula, about the earliest inhabitants of which we know next to nothing. There are two ways to Asia Minor, the one over the Bosphorus and the other through the passes of the Caucasus, and the Aryans may well have utilised both. Finally, the south-eastern tribes probably spread themselves gradually over west Turkestan, and, after evolving the primitive Indo-Iranian dialect, eventually colonised Persia and Hindostan, where their speech developed into its final forms. On this hypothesis, the notion that the Celts and the Teutons migrated from about Pamir and the Hindoo-Koosh is as far from the truth as the supposition that the Indo-Iranians migrated from

Scandinavia. It supposes that the blond long-heads, in what may be called their nascent Aryan stage, that is before their dialects had taken on the full Aryan characteristics, were spread over a wide region which is, conventionally, European; but which, from the point of view of the physical geographer, is rather to be regarded as a continuation of Asia. Moreover, it is quite possible and even probable, that the blond long-heads may have arrived in Turkestan before their language had reached, or at any rate passed beyond, the stage of primitive Aryan; and that the whole process of differentiation into Indo-Iranian took place during the long ages of their residence in the basin of the Oxus. Thus, the question whether the seat of the primitive Aryans was in Europe, or in Asia, becomes very much a debate about geographical terminology.

The foregoing arguments in favour of Latham's "Sarmatian hypothesis" have been based upon data which lie within the ken of history or may be surely concluded by reasoning backwards from the present state of things. But, thanks to the investigations of the pre-historic archæologists and anthropologists during the last half-century, a vast mass of positive evidence respecting the distribution and the condition of mankind in the long interval between the dawn of history and the commencement of the recent epoch has been brought to light.

During this period, there is evidence that men existed in all those regions of Europe which have yet been properly examined; and such of their bony remains as have been discovered exhibit no less diversity of stature and cranial conformation than at present. There are tall and short men; long-skulled and broad-skulled men; and it is probably safe to conclude that the present contrast of blonds and brunets existed among them when they were in the flesh. Moreover it has become clear that, everywhere, the oldest of these people were in the so-called neolithic stage of civilisation. That is to say, they not merely used stone implements which were chipped into shape, but they also employed tools and weapons brought to an edge by grinding. At first they know little or nothing of the use of metals; they possess domestic animals and cultivated plants and live in houses of simple construction.

In some parts of Europe little advance seems to have been made, even down to historical times. But in Britain, France, Scandinavia, Germany, Western Russia, Switzerland, Austria, the plain of the Po, very probably also in the Balkan peninsula, culture gradually advanced until a relatively high degree of civilisation was attained. The initial impulse in this course of progress appears to have been given by the discovery that metal is a better material for tools and weapons than stone. In the early days of pre-historic archæology, Nilsson showed that, in the interments of the middle age, bronze largely took the place of stone, and that, only in the latest, was iron substituted for bronze. Thus arose the generalisation of the occurrence of a regular succession of stages of culture, which were somewhat unfortunately denominated the "ages" of stone, bronze, and iron. For a long time after this order of succession in the same locality (which, it was sometimes forgotten, has nothing to do with chronological contemporaneity in different localities) was made out, the change from stone to bronze was ascribed to foreign, and, of course, Eastern influences. There were the ubiquitous Phœnician traders and the immigrant Aryans from the Hindoo-Koosh, ready to hand. But further investigation has proved[1] for various parts of Europe and made it probable for others, that though the old order of succession is correct it is incomplete, and that a copper stage must be interpolated between the neolithic and the bronze stages. Bronze is an artificial product, the formation of which implies a knowledge of copper; and it is certain that copper was, at a very early period, smelted out of the native ores, by the people of central Europe who used it. When they learned that the hard-

---

[1] "Proved" is perhaps too strong a word. But the evidence set forth by Dr. Much (*Die Kupferzeit in Europa*, 1886) in favour of a copper stage of culture among the inhabitants of the pile-dwellings is very weighty.

ness and toughness of their metal were immensely improved by alloying it with a small quantity of tin, they forsook copper for bronze, and gradually attained a wonderful skill in bronze-work. Finally, some of the European people became acquainted with iron, and its superior qualities drove out bronze, as bronze had driven out stone, from use in the manufacture of implements and weapons of the best class. But the process of substitution of copper and bronze for stone was gradual, and, for common purposes, stone remained in use long after the introduction of metals.

The pile-dwellings of Switzerland have yielded an unbroken archæological record of these changes. Those of eastern Switzerland ceased to exist soon after the appearance of metals, but in those of the Lakes of Neuchatel and Bienne the history is continued through the stage of bronze to the beginning of that of iron. And in all this long series of remains, which lay bare the minutest details of the life of the pile-dwellers, from the neolithic to the perfected bronze stage, there is no indication of any disturbance such as must have been caused by foreign invasion; and such as was produced by intruders, shortly after the iron stage was reached. Undoubtedly the constructors of the pile-dwellings must have received foreign influences through the channel of trade, and may have received them by the slow immigration of other races. Their amber, their jade, and their tin show that they had commercial intercourse with somewhat distant regions. The amber, however, takes us no further than the Baltic; and it is now known that jade is to be had within the boundaries of Europe, while tin lay no further off than north Italy. An argument in favour of oriental influence has been based upon the characters of certain of the cultivated plants and domesticated animals. But even that argument does not necessarily take us beyond the limits of south-eastern Europe; and it needs reconsideration in view of the changes of physical geography and of climate to which I have drawn attention.

In connection with this question there is another important series of facts to be taken into consideration. When, in the seventeenth century, the Russians advanced beyond the Ural and began to occupy Siberia, they found that the majority of the natives used implements of stone and bone. Only a few possessed tools or weapons of iron, which had reached them by way of commerce; the Ostiaks and the Tartars of Tom, alone, extracted their iron from the ore. It was not until the invaders reached the Lena, in the far east, that they met with skilful smiths among the Jakuts,[1] who manufactured knives, axes, lances, battle-axes, and leather jerkins studded with iron;

---

[1] Andree, *Die Metalle bei den Naturvölkern* (p. 114). It is interesting to note that the Jakuts have always been pastoral nomads, formerly shepherds, now horse-breeders, and that they

and among the Tunguses and Lamuts, who had learned from the Jakuts.

But there is an older chapter of Siberian history which was closed in the seventeenth century, as that of the people of the pile-dwellings of Switzerland had ended when the Romans entered Helvetia. Multitudes of sepulchral tumuli, termed like those of European Russia, "kurgans," are scattered over the north Asiatic plains, and are especially agglomerated about the upper waters of the Jenisei. Some are modern, while others, extremely ancient, are attributed to a quasi-mythical people, the Tschudes. These Tschudish kurgans abound in copper and gold articles of use and luxury, but contain neither bronze nor iron. The Tschudes procured their copper and their gold from the metalliferous rocks of the Ural and the Altai; and their old shafts, adits, and rubbish heaps led the Russians to the rediscovery of the forgotten stores of wealth. The race to which the Tschudes belonged and the age of the works which testify to their former existence, are alike unknown. But seeing that a rumour of them appears to have reached Herodotus, while, on the other hand, the pile-dwelling civilisation of Switzerland may perhaps come down as late as the fifth century B.C., the

continue to work their iron in the primitive fashion; as the argument that metallurgic skill implies settled agricultural life not unfrequently makes its appearance.

possibility that a knowledge of the technical value of copper may have travelled from Siberia westward must not be overlooked. If the idea of turning metals to account must needs be Asiatic, it may be north Asiatic just as well as south Asiatic. In the total absence of trustworthy chronological and anthropological data, speculation may run wild.

The oldest civilisations for which we have an, even approximately, accurate chronology are those of the valleys of the Nile and of the Euphrates. Here, culture seems to have attained a degree of perfection, at least as high as that of the bronze stage, six thousand years ago. But before the intermediation of Etruscan, Phœnician, and Greek traders, there is no evidence that they exerted any serious influence upon Europe or northern Asia. As to the old civilisation of Mesopotamia, what is to be said until something definite is known about the racial characters of its originators, the Accadians? As matters stand, they are just as likely to have been a group of the same race as the Egyptians, or the Dravidians, as anything else. And considering that their culture developed in the extreme south of the Euphrates valley, it is difficult to imagine that its influence could have spread to northern Eurasia except by the Phœnician (and Carian?) intermediation which was undoubtedly operative in comparatively late times.

Are we then to bring down the discovery of the use of copper in Switzerland to, at earliest, 1500 B.C., and to put it down to Phœnician hints? But why copper? At that time the Phœnicians must have been familiar with the use of bronze. And if, on the other hand, the northern Eurasiatics had got as far as copper, by the help of their own ingenuity, why deny them the capacity to make the further step to bronze? Carry back the borrowing system as far as we may, in the end we must needs come to some man or men from whom the novel idea started, and who after many trials and errors gave it practical shape. And there really is no ground in the nature of things for supposing that such men of practical genius may not have turned up, independently, in more races than one.

The capacity of the population of Europe for independent progress while in the copper and early bronze stage—the "palæo-metallic" stage, as it might be called—appears to me to be demonstrated in a remarkable manner by the remains of their architecture. From the crannog to the elaborate pile-dwelling, and from the rudest enclosure to the complex fortification of the terramare, there is an advance which is obviously a native product. So with the sepulchral constructions; the stone cist, with or without a preservative or memorial cairn, grows into the chambered graves lodged in tumuli; into such megalithic edifices as the dromic vaults of Maes How and New Grange; to culminate in the finished masonry of the tombs of Mycenæ, constructed on exactly the same plan. Can any one look at the varied series of forms which lie between the primitive five or six flat stones fitted together into a mere box, and such a building as Maes How, and yet imagine that the latter is the result of foreign tuition? But the men who built Maes How, without metal tools, could certainly have built the so-called "treasure-house" of Mycenæ, with them.

If these old men of the sea, the heights of Hindoo-Koosh-Pamir and the plain of Shinar, had been less firmly seated upon the shoulders of anthropologists, I think they would long since have seen that it is at least possible that the early civilisation of Europe is of indigenous growth; and that, so far as the evidence at present accumulated goes, the neolithic culture may have attained its full development, copper may have gradually come into use, and bronze may have succeeded copper, without foreign intervention.

So far as I am aware, every raw material employed in Europe up to the palæo-metallic stage, is to be found within the limits of Europe; and there is no proof that the old races of domesticated animals and plants could not have been developed within these limits. If any one chose to maintain, that the use of bronze in Europe originated among the inhabitants of Etruria and radiated thence, along the already established lines of traffic to all parts of Europe, I do not see that his contention could be upset. It would be hard to prove either that the primitive Etruscans could not have discovered the way to manufacture bronze, or that they did not discover it and become a great mercantile people in consequence, before Phœnician commerce had reached the remote shores of the Tyrrhene Sea.

Can it be safely concluded that the palæometallic culture which we have been considering was the appanage of any one of the western Eurasiatic races rather than another? Did it arise and develop among the brunet or the blond long-heads, or among the brunet short-heads? I do not think there are any means of answering these questions, positively, at present. Schrader has pointed out that the state of culture of the primitive Aryans, deduced from philological data, closely corresponds with that which obtained among the pile-dwellers in the neolithic stage. But the resemblance of the early stages of civilisation among the most different and widely separated races of mankind, should warn us that archæology is no more a sure guide in questions of race than philology.

With respect to the osteological characters of

the people of the Swiss pile-dwellings information is as yet scanty. So far as the present evidence goes, they appear to have comprised both broadheads and long-heads of moderate stature.[1] In France, England, and Germany, both long and broad skulls are found in tumuli belonging to the neolithic stage. In some parts of England the long skulls, and in others the broad skulls, accompany the higher stature. In the Scandinavian peninsula, nine-tenths of the neolithic people are decided long-heads: in Denmark, there is a much larger proportion of broad-heads.

In view of all the facts known to me (which cannot be stated in greater detail in this place), I am disposed to think that the blond long-heads, the brunet long-heads, and the brunet broadheads have existed on the continent of Europe throughout the Recent period: that only the former two at first inhabited our islands; but that a mixed race of tall broad-heads, like some of the Blackforesters of the present day, so excellently described by Ecker, migrated from the continent and formed that tall contingent of the population

[1] Professor Virchow has guardedly expressed the opinion that the oldest inhabitants of the Swiss pile-dwellings were broadheads, and that later on (commencing before the bronze stage) there was a gradual infusion of long-heads among them. (*Zeitschrift für Ethnologie.* xvii., 1885). There is independent evidence of the existence of broad-heads in the Cevennes during the neolithic period, and I should be disposed to think that this opinion may well be correct; but the examination of the evidence on which it is, at present, based does not lead me to feel very confident about it.

which has been identified (rightly or wrongly) with the Belgæ by Thurnam and which seems to have subsequently lost itself among the predominant brunet and blond long-heads.

I do not think there is anything to warrant the conclusion that the palæo-metallic culture of Europe took its origin among the blond long-head (or supposed Aryan) race; or that the people of the Swiss pile-dwellings belonged to that race. The long-heads among them may just as likely have been brunets. In north-eastern Italy there is clear evidence of the superposition of at least four stages of culture, in which that of the copper and bronze using terramare people comes second; a stage marked by Etruscan domination occupies the third place; and that is followed by the stage which appertains to the Gauls, with their long swords and other characteristic iron work. In western Switzerland, on the other hand, at La Téne, and elsewhere, similar relics show that the Gauls followed upon the latest population of the pile-dwellings among whom traces of Etruscan influence (though not of dominion) are to be found. Helbig supposes the terramare people to have been Greco-Latin-speaking Pelasgi, and consequently Aryan. But we cannot suppose the people of the pile-dwellings of Switzerland to have been speakers of primitive Greco-Latin (if ever there was such a language). And if the Gauls were the first speakers of Celtic who got into Switzerland, what Aryan language can the people of the pile-dwellings have spoken?[1]

As I have already mentioned, there is not the least doubt that man existed in north-western Europe during the Pleistocene or Quaternary epoch. It is not only certain that men were contemporaries of the mammoth, the hairy rhinoceros, the reindeer, the cave bear, and other great carnivora, in England and in France, but a great deal has been ascertained about the modes of life of our predecessors. They were savage hunters, who took advantage of such natural shelters as overhanging rocks and caves, and perhaps built themselves rough wigwams; but who had no domestic animals and have left no sign that they cultivated plants. In many localities there is evidence that a very considerable interval—the so-called *hiatus*—intervened between the time when the Quaternary or palæolithic men occupied particular caves and river basins and the accumulation of the debris left by their neolithic successors. And, in spite of all the warnings against negative evidence afforded by the history of geology, some have very positively asserted that this means a complete break between the Quater-

[1] See Dr. Munro's excellent work, *The Lake Dwellings of Europe*, for La Téne. Readers of Professor Rhys' recent articles (*Scottish Review*, 1890) may suggest that the pile-dwelling people spoke the Gaedhelic form of Celtic, and the Gauls the Brythonic form.

nary and the Recent populations—that the Quaternary population followed the retreating ice northwards and left behind them a desert which remained unpeopled for ages. Other high authorities, on the contrary, have maintained that the races of men who now inhabit Europe may all be traced back to the Great Ice Age. When a conflict of opinion of this kind obtains among reasonable and instructed men, it is generally a safe conclusion that the evidence for neither view is worth much. Certainly that is the result of my own cogitations with regard to both the hiatus doctrine (in its extreme form) and its opposite—though I think the latter by much the more likely to turn out right. But I hesitate to adopt it on the evidence which has been obtained up to this time.

No doubt, human bones and skulls of various types have been discovered in close proximity to palæolithic implements and to skeletons of quaternary quadrupeds; no doubt, if the bones and skulls in question were not human, their contemporaneity would hardly have been questioned. But, since they are human, the demand for further evidence really need not be ascribed to mere conservative prejudice. Because the human biped differs from all other bipeds and quadrupeds, in the tendency to put his dead out of sight in various ways; commonly by burial. It is a habit worthy of all respect in itself, but generative of subtle traps and grievous pitfalls for the unwary investigator of human palæontology. For it may easily happen, that the bones of him that "died o' Wednesday," may thus come to lie alongside the bones of animals that were extinct thousands of years before that Wednesday; and yet the interment may have been effected so many thousands of years ago that no outward sign betrays the difference in date. In all investigations of this kind, the most careful and critical study of the circumstances is needful if the results are to be accepted as perfectly trustworthy.

In the case of the remains found in a cave of the valley of the Neander, near Düsseldorf, half a century ago—the characters of which gave rise to a vast amount of discussion at that time and subsequently—the circumstances of the discovery were but vaguely known. The skeleton was met with in a deposit, the loess, which is known to be of quaternary age; there was no evidence to show how it came there. Consequently, not only was its exact age justly and properly declared to be a matter of doubt; but those who, on scientific or other grounds, were inclined to minimise its importance could put forth plausible speculations about its nature which do not look so well under the light thrown by a more advanced science of Anthropology. It could be and it was suggested that the Neanderthal skeleton was that of a strayed idiot; that the characters of the skull were the result of early synostosis or of late gout; and, in fact, any stick was good enough to beat the dog withal.

As some writings of mine on the subject led to my occupation of a prominent position among the belaboured dogs of that day, I have taken a mild interest in watching the gradual rehabilitation of my old friend of the Neanderthal among normal men, which has been going on of late years. It has come to be generally admitted that his remarkable cranium is no more than a strongly-marked example of a type which occurs, not only among other prehistoric men, but is met with, sporadically, among the moderns; and that, after all, I was not so wrong as I ought to have been, when I indicated such points of similarity among the skulls found in our river-beds and among the native races of Australia.[1] However, doubts still clung about the geological age of the various deposits in which skulls of the Neanderthal type were subsequently found; and it was not until the year 1886 that two highly-competent observers, Messrs. Fraipont and Lohest, the one an anatomist, the other a geologist, furnished us with evidence such as will bear severe criticism. At the mouth of a cave in the commune of Spy, in the Belgian province of Namur, Messrs. Fraipont and Lohest discovered two skeletons of the Neanderthal type; and the elaborate account of their investigations which they have published appears to me to leave

[1] See p. 202 of this volume.

little room for doubt that the men of Spy fabricated the palæolithic implements, and were the contemporaries of the characteristic quaternary quadrupeds, found with them. The anatomical characters of the skeletons bear out conclusions which are not flattering to the appearance of the owners. They were short of stature but powerfully built, with strong, curiously-curved, thigh-bones, the lower ends of which are so fashioned that they must have walked with a bend at the knees. Their long depressed skulls had very strong brow ridges; their lower jaws, of brutal depth and solidity, sloped away from the teeth downwards and backwards, in consequence of the absence of that especially characteristic feature of the higher type of man, the chin prominence. Thus these skulls are not only eminently "Neanderthaloid," but they supply the proof that the parts wanting in the original specimen harmonised in lowness of type with the rest.

After a very full discussion of the anatomical characters of these skulls, M. Fraipont says:

To sum up, we consider ourselves to be in a position to say that, having regard merely to the anatomical structure of the man of Spy, he possessed a greater number of pithecoid characters than any other race of mankind.[1]

And after enumerating these he continues:

The other and much more numerous characters of the skull, of

---

[1] Fraipont et Lohest. "La Race humaine de Néanderthal, ou de Canstatt, en Belgique," *Archives de Biologie*, 1886.

the trunk, and of the limbs seem to be all human. Between the man of Spy and an existing anthropoid ape there lies an abyss.

Now that is pleasant reading for me, because, in 1863, I committed myself to the assertion that the Neanderthal skull was "the most pithecoid of human crania yet discovered," yet that " in no sense can the Neanderthal bones be regarded as the remains of a human being intermediate between men and apes" [1] and " that the fossil remains of Man hitherto discovered do not seem to me to take us appreciably nearer to that lower pithecoid form, by the modification of which he has, probably, become what he is." [2]

As the evidence stood seven and twenty years ago, in fact, it would have been imprudent to assume that the Neanderthal skull was anything but a case of sporadic reversion. But, in my anxiety not to overstate my case, I understated it. The Neanderthaloid race is "appreciably nearer," though the approximation is but slight. In the words of M. Fraipont:

> The distance which separates the man of Spy from the modern anthropoid ape is undoubtedly enormous; between the man of Spy and the *Dryopithecus* it is a little less. But we must be permitted to point out that if the man of the later quaternary age is the stock whence existing races have sprung, he has travelled a very great way.
>
> From the data now obtained, it is permissible to believe that

---

[1] See p. 205 *supra*.   [2] *Ibid*, p. 208.

we shall be able to pursue the ancestral type of men and the anthropoid apes still further, perhaps as far as the eocene and even beyond.[1]

These conclusions hold good whatever the age of the men of Spy; but they possess a peculiar interest if we admit, as I think on the evidence must be admitted, that these human fossils are of pleistocene age. For, after all due limitations, they give us some, however dim, insight into the rate of evolution of the human species, and indicate that it has not taken place at a much faster or slower pace than that of other mammalia. And if that is so, we are warranted in the supposition that the genus *Homo*, if not the species which the courtesy or the irony of naturalists has dubbed *sapiens*, was represented in pliocene, or even in miocene times. But I do not know by what osteological peculiarities it could be determined whether the pliocene, or miocene, man was sufficiently sapient to speak or not;[2] and whether, or not, he answered to the definition "rational animal" in any higher sense than a dog or an ape does.

There is no reason to suppose that the genus

[1] "Where, then, must we look for primæval Man? Was the oldest *Homo sapiens*, pliocene or miocene, or yet more ancient? In still older strata do the fossilised bones of an Ape more anthropoid or a Man more pithecoid than any yet known await the researches of some unborn palæontologist?"—P. 208 *supra*.

[2] I am perplexed by the importance attached by some to the presence or absence of the so-called "genial" elevations. Does any one suppose that the existence of the genio-hyo-glossus muscle, which plays so large a part in the movements of the tongue, depends on that of these elevations?

Homo was confined to Europe in the pleistocene age; it is much more probable that this, like other mammalian genera of that period, was spread over a large extent of the surface of the globe. At that time, in fact, the climate of regions nearer the equator must have been far more favourable to the human species; and it is possible that, under such conditions, it may have attained a higher development than in the north. As to where the genus *Homo* originated, it is impossible to form even a probable guess. During the miocene epoch, one region of the present temperate zones would serve as well as another. The elder Agassiz long ago tried to prove that the well-marked areas of geographical distribution of mammals have their special kinds of men; and, though this doctrine cannot be made good to the extent which Agassiz maintained; yet the limitation of the Australian type to New Holland,[1] the approximate restriction of the negro type to Ultra-Saharal Africa, and the peculiar character of the population of Central and South America, are facts which bear strongly in favour of the conclusion that the causes which have influenced the distribution of mammals in general, have powerfully affected that of man.

Let it be supposed that the human remains from the caves of the Neanderthal and of Spy

[[1] Unless I am right in extending it to Hindostan and even further west.—1894.]

represent the race, or one of the races, of men who inhabited Europe in the quaternary epoch, can any connection be traced between it and existing races? That is to say, do any of them exhibit characters approximating those of the Spy men or other examples of the Neanderthaloid race? Put in the latter form, I think that the question may be safely answered in the affirmative. Skulls do occasionally approach the Neanderthaloid type, among both the brunet and the blond long-head races. For the former, I pointed out the resemblance, long ago, in some of the Irish river-bed skulls. For the latter, evidence of various kinds may be adduced; but I prefer to cite the authority of one of the most accomplished and cautious of living anthropologists. Professor Virchow was led, by historical considerations, to think that the Teutonic type, if it still remained pure and undefiled anywhere, should be discoverable among the Frisians, in their ancient island homes on the North German coast, remote from the great movements of nations. In their tall stature and blond complexion the Frisians fulfilled expectation; but their skulls differed in some respects from those of the neighbouring blond long-heads. The depression, or flattening (accompanied by a slight increase in breadth), which occurs occasionally among the latter, is regular and characteristic among the Frisians; and, in other respects, the Frisian skull unmistakably approaches the Neanderthal and Spy type.[1] The fact that this resemblance exists is of none the less importance because the proper interpretation of it is not yet clear. It may be taken to be a pretty sure indication of the physiological continuity of the blond long-heads with the pleistocene Neanderthaloid men. But this continuity may have been brought about in two ways. The blond long-heads may exhibit one of the lines of evolution of the men of the Neanderthaloid type. Or, the Frisians may be the result of the admixture of the blond long-heads with Neanderthaloid men; whose remains have been found at Canstatt and at Gibraltar, as well as at Spy and in the valley of the Neander; and who, therefore, seem, at one time, to have occupied a considerable area in Western Europe. The same alternatives present themselves when Neanderthaloid characters appear in skulls of other races. If these characters belong to a stage in the development of the human species, antecedent to the differentiation of any of the existing races, we may expect to find them in the lowest of these races, all over the world, and in the early stages of all races. I have already referred to the remarkable similarity of the skulls of certain tribes of native Australians to the

[1] Virchow *Beiträge zur physischen Anthropologie der Deutschen* (*Abh. der Königlichen Akademie der Wissenschaften zu Berlin*, 1876). See particularly p. 238 for the full recognition of the Neanderthaloid characters of Frisian skulls and of the ethnological significance of the similarity.

Neanderthal skull; and I may add, that the wide differences in height between the skulls of different tribes of Australians afford a parallel to the differences in altitude between the skulls of the men of Spy and those of the grave rows of North Germany. Neanderthaloid features are to be met with, not only in ancient long skulls; those of the ancient broad-headed people entombed at Borreby in Denmark have been often noted.

Reckoned by centuries, the remoteness of the quaternary, or pleistocene, age from our own is immense, and it is difficult to form an adequate notion of its duration. Undoubtedly there is an abysmal difference between the Neanderthaloid race and the comely living specimens of the blond long-heads with whom we are familiar. But the abyss of time between the period at which North Europe was first covered with ice, when savages pursued mammoths and scratched their portraits with sharp stones in central France, and the present day, ever widens as we learn more about the events which bridge it. And, if the differences between the Neanderthaloid men and ourselves could be divided into as many parts as that time contains centuries, the progress from part to part would probably be almost imperceptible.

END OF VOL. VII

www.ingramcontent.com/pod-product-compliance
Lightning Source LLC
Chambersburg PA
CBHW030323240426
43673CB00040B/1262